AN
INCONVENIENT
TRUTH

CANEY FORK RIVER,
CARTHAGE, TN, 2006.
PHOTOGRAPH BY TIPPER GORE

Published by

33 E. Minor Street
Emmaus, PA 18098
www.rodale.com

Produced by

MELCHER
MEDIA

124 West 13th Street
New York, NY 10011
www.melcher.com

Rodale books may be purchased for business or promotional use or
for special sales. For information, please write to:
Special Markets Department, Rodale, Inc.,
733 Third Avenue, New York, NY 10017

09 08 07 06/10 9 8 7 6 5 4

Printed in the USA
Design: mgmt. design

ISBN-13: 978-1-59486-567-1
ISBN-10: 1-59486-567-1
Library of Congress Control Number: 2006926537

AN INCONVENIENT TRUTH

THE PLANETARY EMERGENCY OF GLOBAL WARMING AND WHAT WE CAN DO ABOUT IT

AL GORE

AL AND TIPPER GORE, ONE
MONTH BEFORE THE BIRTH OF
THEIR FIRST CHILD, KARENNA,
ON THE CANEY FORK RIVER,
CARTHAGE, TN, 1973

For my beloved wife and partner, Tipper,
who has been with me for the entire journey

Introduction

Some experiences are so intense while they are happening that time seems to stop altogether. When it begins again and our lives resume their normal course, those intense experiences remain vivid, refusing to stay in the past, remaining always and forever with us.

Seventeen years ago my youngest child was badly—almost fatally—injured. This is a story I have told before, but its meaning for me continues to change and to deepen.

That is also true of the story I have tried to tell for many years about the global environment. It was during that interlude 17 years ago when I started writing my first book, *Earth in the Balance*. It was because of my son's accident and the way it abruptly interrupted the flow of my days and hours that I began to rethink everything, especially what my priorities had been. Thankfully, my son has long since recovered completely. But it was during that traumatic period that I made at least two enduring changes: I vowed always to put my family first, and I also vowed to make the climate crisis the top priority of my professional life.

Unfortunately, in the intervening years, time has not stood still for the global environment. The pace of destruction has worsened and the urgent need for a response has grown more acute.

The fundamental outline of the climate crisis story is much the same now as it was then. The relationship between human civilization and the Earth has been utterly transformed by a combination of factors, including the population explosion, the technological revolution, and a willingness to ignore the future consequences of our present actions. The underlying reality is that we are colliding with the planet's ecological system, and its most vulnerable components are crumbling as a result.

I have learned much more about this issue over the years. I have read and listened to the world's leading scientists, who have offered increasingly dire warnings. I have watched with growing concern as the crisis gathers strength even more rapidly than anyone expected.

In every corner of the globe—on land and in water, in melting ice and disappearing snow, during heat waves and droughts, in the eyes of hurricanes and in the tears of refugees—the world is witnessing mounting and undeniable evidence that nature's cycles are profoundly changing.

I have learned that, beyond death and taxes, there is at least one absolutely indisputable fact: Not only does human-caused global warming exist, but it is also growing more and more dangerous, and at a pace that has now made it a planetary emergency.

Part of what I have learned over the last 14 years has resulted from changes in my personal circumstances as well. Since 1992 much has happened in my life. Our children have all grown up, and our two oldest daughters have married. Tipper and I now have two grandchildren. Both of my parents have died, as has Tipper's mother.

And less than a year after *Earth in the Balance* was published, I was elected vice president—ultimately serving for eight years. I had the opportunity, as a member of the Clinton-Gore administration, to pursue an ambitious agenda of new policies addressing the climate crisis.

At that time I discovered, firsthand, how fiercely Congress would resist the changes we were urging them to make, and I watched with growing dismay as the opposition got much, much worse after the takeover of Congress in 1994 by the Republican party and its newly aggressive conservative leaders.

I organized and held countless events to spread public awareness about the climate crisis, and to build more public support for congressional action. I also learned numerous lessons about the significant changes in recent decades in the nature and quality of America's "conversation of democracy." Specifically, that entertainment values have transformed what we used to call news, and individuals with independent voices are routinely shut out of the public discourse.

In 1997 I helped achieve a breakthrough at the negotiations in Kyoto, Japan, where the world drafted a groundbreaking treaty whose goal is to control global warming pollution. But then I came home and faced an uphill battle to gain support for the treaty in the U.S. Senate.

In 2000 I ran for president. It was a long and hard-fought campaign that was ended by a 5–4 decision in the Supreme Court to halt the counting of votes in the key state of Florida. This was a hard blow.

I then watched George W. Bush get sworn in as president. In his very first week in office, President Bush reversed a campaign pledge to regulate CO_2 emissions—a pledge that had helped persuade many voters that he was genuinely concerned about matters relating to the environment.

Soon after the election, it became clear that the Bush-Cheney administration was determined to block any policies designed to help limit global-warming pollution. They launched an all-out effort to roll back, weaken, and—wherever possible—completely eliminate existing laws and regulations. Indeed, they even abandoned Bush's pre-election rhetoric about global warming, announcing that, in the president's opinion, global warming wasn't a problem at all.

As the new administration was getting underway, I had to begin making decisions about what I would do in my own life. After all, I was now out of a job. This certainly wasn't an easy time, but it did offer me the chance to make a fresh start—to step back and think about where I should direct my energies.

I began teaching courses at two colleges in Tennessee, and, along with Tipper, published two books about the American family. We moved to Nashville and bought a house less than an hour's drive from our farm in Carthage. I entered the business world and eventually started two new companies. I became an adviser to two already established major high-tech businesses.

I am tremendously excited about these ventures, and feel fortunate to have found ways to make a living while simultaneously moving the world—at least a little—in the right direction.

With my partner Joel Hyatt I started Current TV, a news and information cable and satellite network for young people in their twenties, based on an idea that is, in our present-day society, revolutionary: that viewers themselves can make the programs and in the process participate in the public forum of American democracy. With my partner David Blood I also started Generation Investment Management, a firm devoted to proving that the environment and other sustainability factors can be fully integrated into the mainstream investment process in a way that enhances profitability for our clients, while encouraging businesses to operate more sustainably.

At first, I thought I might run for president again, but over the last several years I have discovered that there are other ways to serve, and that I am really enjoying them.

I am also determined to continue to make speeches on public policy, and—as I have at almost every crossroads moment in my life—to make the global environment my central focus.

Since my childhood summers on our family's farm in Tennessee, when I first learned from my father about taking care of the land, I have been deeply interested in learning more about threats to the environment. I grew up half in the city and half in the country, and the half I loved most was on our farm. Since my mother read to my sister and me from Rachel Carson's classic book, *Silent Spring*, and especially since I was first introduced to the idea of global warming by my college professor Roger Revelle, I have always tried to deepen my own understanding of the human impact on nature, and in my public service I have tried to implement policies that would ameliorate—and eventually eliminate—that harmful impact.

During the Clinton-Gore years we accomplished a lot in terms of environmental issues, even though, with the hostile Republican Congress, we fell short of all that was needed. Since the change in administrations, I have watched with growing concern as our forward progress has been almost completely reversed.

After the 2000 election, one of the things I decided to do was to start giving my slide show on global warming again. I had first put it together at the same time I began writing *Earth in the Balance*, and over the years I have added to it and steadily improved it to the point where I think it makes a compelling case, at least for most audiences, that humans are the cause of most of the global warming that is taking place, and that unless we take quick action the consequences for our planetary home could become irreversible.

For the last six years, I have been traveling around the world, sharing the information I have compiled with anyone who would listen. I have traveled to colleges, to small towns and big cities. More and more, I have begun to feel that I am changing minds, but it is a slow process.

I was giving my presentation to a group in Los Angeles one evening in the spring of 2005, and afterward, several people came up to me and suggested I consider making a film about global warming. This particular audience included some well-known figures in the entertainment industry, including environmental activist Laurie David and film producer Lawrence Bender, and so I knew their intentions were serious. But I had no idea how my slide show could possibly translate to film. They requested a second meeting and introduced me to Jeff Skoll, founder and CEO of Participant Productions, who offered to finance the movie, and to a highly talented film veteran, Davis Guggenheim, who expressed interest in directing it. Later, Scott Burns joined the production team and Lesley Chilcott became the coproducer and legendary "trail boss."

My principal concern in all this was that the translation of the slide show into a film not sacrifice the central role of science for entertainment's sake. But the more I talked with this extraordinary group, and felt their deep commitment to exactly the same goals I was pursuing, the more convinced I became that the movie was a good idea. If I wanted to reach the maximum number of people quickly, and not just continue talking to a few hundred people a night, a movie was the way to do it. That film, also titled *An Inconvenient Truth*, has now been made, and I am really excited about it.

But the idea for a book on the climate crisis actually came first. It was Tipper who first suggested that I put together a new kind of book with pictures and graphics to make the whole message easier to follow, combining many elements from my slide show with all of the new original material I have compiled over the last few years.

Tipper and I are, by the way, giving 100% of whatever profits come to us from the book—and from the movie—to a non-profit, bipartisan effort to move public opinion in the United States to support bold action to confront global warming.

After more than thirty years as a student of the climate crisis, I have a lot to share. I have tried to tell this story in a way that will interest all kinds of readers. My hope is that those who read the book and see the film will begin to feel, as I have for a long time, that global warming is not just about science and that it is not just a political issue. It is really a moral issue.

Although it is true that politics at times must play a crucial role in solving this problem, this is the kind of challenge that ought to completely transcend partisanship. So whether you are a Democrat or a Republican, whether you voted for me or not, I very much hope that you will sense that my goal is to share with you both my passion for the Earth and my deep sense of concern for its fate. It is impossible to feel one without the other when you know all the facts.

I also want to convey my strong feeling that what we are facing is not just a cause for alarm, it is paradoxically also a cause for hope. As many know, the Chinese expression for "crisis" consists of two characters side by side 危机. The first is the symbol for "danger," the second the symbol for "opportunity."

The climate crisis is, indeed, extremely dangerous. In fact it is a true planetary emergency. Two thousand scientists, in a hundred countries, working for more than 20 years in the most elaborate and well-organized scientific collaboration in the history of humankind, have forged an exceptionally strong consensus that all the nations on Earth must work together to solve the crisis of global warming.

The voluminous evidence now strongly suggests that unless we act boldly and quickly to deal with the underlying causes of global warming, our world will undergo a string of terrible catastrophes, including more and stronger storms like Hurricane Katrina, in both the Atlantic and the Pacific.

We are melting the North Polar ice cap and virtually all of the mountain glaciers in the world. We are destabilizing the massive mound of ice on Greenland and the equally enormous mass of ice propped up on top of islands in West Antarctica, threatening a worldwide increase in sea levels of as much as 20 feet.

The list of what is now endangered due to global warming also includes the continued stable configuration of ocean and wind currents that has been in place since before the first cities were built almost 10,000 years ago.

We are dumping so much carbon dioxide into the Earth's environment that we have literally changed the relationship between the Earth and the Sun. So much of that CO_2 is being absorbed into the oceans that if we continue at the current rate we will increase the saturation of calcium carbonate to levels that will prevent formation of corals and interfere with the making of shells by any sea creature.

Global warming, along with the cutting and burning of forests and other critical habitats, is causing the loss of living species at a level comparable to the extinction event that wiped out the dinosaurs 65 million years ago. That event was believed to have been caused by a giant asteroid. This time it is not an asteroid colliding with the Earth and wreaking havoc; it is us.

Last year, the national academies of science in the 11 most influential nations came together to jointly call on every nation to "acknowledge that the threat of climate change is clear and increasing" and declare that the "scientific understanding of climate changes is now sufficiently clear to justify nations taking prompt action."

So the message is unmistakably clear. This crisis means "danger!"

Why do our leaders seem not to hear such a clear warning? Is it simply that it is inconvenient for them to hear the truth?

If the truth is unwelcome, it may seem easier just to ignore it.

But we know from bitter experience that the consequences of doing so can be dire.

For example, when we were first warned that the levees were about to break in New Orleans because of Hurricane Katrina, those warnings were ignored. Later, a bipartisan group of members of Congress chaired by Representative Tom Davis (R-VA), chairman of the House Government Reform Committee, said in an official report, "The White House failed to act on the massive amounts of information at its disposal," and that a "blinding lack of situational awareness and disjointed decision-making needlessly compounded and prolonged Katrina's horror."

Today, we are hearing and seeing dire warnings of the worst potential catastrophe in the history of human civilization: a global climate crisis that is deepening and rapidly becoming more dangerous than anything we have ever faced.

And yet these clear warnings are also being met with a "blinding lack of situational awareness"—in this case, by the Congress, as well as the president.

As Martin Luther King Jr. said in a speech not long before his assassination:

"We are now faced with the fact, my friends, that tomorrow is today. We are confronted with the fierce urgency of now. In this unfolding conundrum of life and history, there is such a thing as being too late.

"Procrastination is still the thief of time. Life often leaves us standing bare, naked, and dejected with a lost opportunity. The tide in the affairs of men does not remain at flood—it ebbs. We may cry out desperately for time to pause in her passage, but time is adamant to every plea and rushes on. Over the bleached bones and jumbled

residues of numerous civilizations are written the pathetic words 'Too late.' There is an invisible book of life that faithfully records our vigilance in our neglect. Omar Khayyam is right: 'The moving finger writes, and having writ moves on.'"

But along with the danger we face from global warming, this crisis also brings unprecedented opportunities.

What are the opportunities such a crisis also offers? They include not just new jobs and new profits, though there will be plenty of both, we can build clean engines, we can harness the Sun and the wind; we can stop wasting energy; we can use our planet's plentiful coal resources without heating the planet.

The procrastinators and deniers would have us believe this will be expensive. But in recent years, dozens of companies have cut emissions of heat-trapping gases while saving money. Some of the world's largest companies are moving aggressively to capture the enormous economic opportunities offered by a clean energy future.

But there's something even more precious to be gained if we do the right thing.

The climate crisis also offers us the chance to experience what very few generations in history have had the privilege of knowing: *a generational mission*; the exhilaration of a compelling *moral purpose*; a shared and unifying *cause*; the thrill of being forced by circumstances to put aside the pettiness and conflict that so often stifle the restless human need for transcendence; *the opportunity to rise*.

When we do rise, it will fill our spirits and bind us together. Those who are now suffocating in cynicism and despair will be able to breathe freely. Those who are now suffering from a loss of meaning in their lives will find hope.

When we rise, we will experience an epiphany as we discover that this crisis is not really about politics at all. It is a moral and spiritual challenge.

At stake is the survival of our civilization and the habitability of the Earth. Or, as one eminent scientist put it, the pending question is whether the combination of an opposable thumb and a neocortex is a viable combination on this planet.

The understanding we will gain—about who we really are—will give us the moral capacity to take on other related challenges that are also desperately in need of being redefined as moral imperatives with practical solutions: HIV/AIDS and other pandemics that are ravaging so many; global poverty; the ongoing redistribution of wealth globally from the poor to the wealthy; the ongoing genocide in Darfur; the ongoing famine in Niger and elsewhere; chronic civil wars; the destruction of ocean fisheries; families that don't function; communities that don't commune; the erosion of democracy in America; and the refeudalization of the public forum.

Consider what happened during the crisis of global fascism. At first, even the truth about Hitler was inconvenient. Many in the west hoped the danger would simply go away. They ignored clear warnings and compromised with evil, and waited, hoping for the best.

After the appeasement at Munich, Churchill said: "This is only the first sip, the first foretaste of a bitter cup which will be proffered to us year by year—unless by supreme recovery of moral health and martial vigor, we rise again and take our stand for freedom."

But when England and then America and our allies ultimately rose to meet the threat, together we won two wars simultaneously in Europe and the Pacific.

By the end of that terrible war, we had gained the moral authority and vision to create the Marshall Plan—and convinced taxpayers to pay for it! We had gained the spiritual capacity and wisdom to rebuild Japan and Europe and launch the renewal of the very nations we had just defeated in war, in the process laying the foundation for 50 years of peace and prosperity.

This too is a moral moment, a crossroads. This is not ultimately about any scientific discussion or political dialogue. It is about who we are as human beings. It is about our capacity to transcend our own limitations, to rise to this new occasion. To see with our hearts, as well as our heads, the response that is now called for. This is a moral, ethical and spiritual challenge.

We should not fear this challenge. We should welcome it. We must not wait. In the words of Dr. King, "Tomorrow is today."

I began this introduction with a description of an experience 17 years ago that, for me, stopped time. During that painful period I gained an ability I hadn't had before to feel the preciousness of our connection to our children and the solemnity of our obligation to safeguard their future and protect the Earth we are bequeathing to them.

Imagine with me now that once again, time has stopped—for all of us—and before it starts again, we have the chance to use our moral imaginations and to project ourselves across the expanse of time, 17 years into the future, and share a brief conversation with our children and grandchildren as they are living their lives in the year 2023.

Will they feel bitterness toward us because we failed in our obligation to care for the Earth that is their home and ours? Will the Earth have been irreversibly scarred by us?

Imagine now that they are asking us: "What were you thinking? Didn't you care about our future? Were you really so self-absorbed that you couldn't—or wouldn't—stop the destruction of Earth's environment?"

What would our answer be?

We can answer their questions now by our actions, not merely with our promises. In the process, we can choose a future for which our children will thank us.

This is the first picture most of us ever saw of the Earth from space. It was taken on Christmas Eve, 1968, during the Apollo 8 mission, the first of the Apollo missions that left the confines of near-Earth orbit and circled the Moon scouting for landing sites before Apollo 11 touched down the following summer.

The vessel went around the far side of the Moon and lost radio contact, as expected. Inevitably, even though everyone understood the reason for the protracted silence, it was a time of great suspense. Then, as radio contact was reestablished, the crew looked up and saw this spectacular sight.

While the crew watched the Earth emerging from the dark void of space, the mission commander, Frank Borman, read from the book of Genesis: "In the beginning God created the Heavens and the Earth."

One of the astronauts aboard, a rookie named Bill Anders, snapped this picture, and it became known as *Earth Rise*. The image exploded into the consciousness of humankind. In fact, within two years of this picture being taken, the modern environmental movement was born. The Clean Air Act, the Clean Water Act, the Natural Environmental Policy Act, and the first Earth Day all came about within a few years of this picture being seen for the first time.

The day after it was taken, on Christmas Day, 1968, Archibald MacLeish wrote:

"To see the Earth as it truly is, small and blue and beautiful in that eternal silence where it floats, is to see ourselves as riders on the Earth together, brothers on that bright loveliness in the eternal cold—brothers who know now that they are truly brothers."

14

This is the last picture of our planet
taken by a human being from space.
It was taken in December 1972 during
the Apollo 17 mission—the last Apollo
mission—from a point halfway between
the Earth and the Moon.

What makes this image so extraordinary
is that it's the only photo we have of the
Earth from space taken when the Sun
was directly behind the spacecraft.

Just as eclipses of the Sun occur only
on those rare occasions when the Earth
and the Sun and the Moon are positioned
along a straight line, this was the only
time during the four-year series of Apollo
missions when the sun was lined up almost directly behind the Moon while the
spacecraft was making its journey. So the
Earth, instead of being partly shrouded
in darkness, appears fully illuminated.

For this reason, this image has become
the most commonly published photograph in all of history. No other image
comes close. In fact, 99 times out of 100,
when you see a picture of Earth, this is
the picture you are seeing.

These magical images of Earth were created by a friend of mine, Tom Van Sant, who went through 3,000 satellite images taken over a three-year period and carefully selected the ones showing a cloud-free view of the Earth's surface. He then digitally stitched together the images to create a composite view of the entire planet, in which all of its surfaces are clearly visible.

Because Van Sant's images are of a globe, the only way to see every part of the Earth simultaneously is to spread out the images in a flat picture, called a projection. Any projection inevitably causes some distortion in the shape and size of the continents, especially Antarctica and the area around the North Pole. But the imagery portrayed here is based on the 3,000 photographs that Van Sant has used to make all of his other stunning views of the Earth.

I am sharing this image to illustrate one of two stories about teachers I once had. The first, my sixth grade teacher, taught geography by pulling down a map of the world in front of a blackboard. One of my classmates raised his hand, pointed to the east coast of South America and the west coast of Africa and asked, "Did they ever fit together?" And the teacher responded, "Of course not! That's the most ridiculous thing I've ever heard!"

That sixth-grade teacher had an assumption in his mind that he didn't bother to question: Continents are so big, obviously they don't move.

As we now know, they did move. At one time they fit together, then moved apart millions of years ago. And they're still moving.

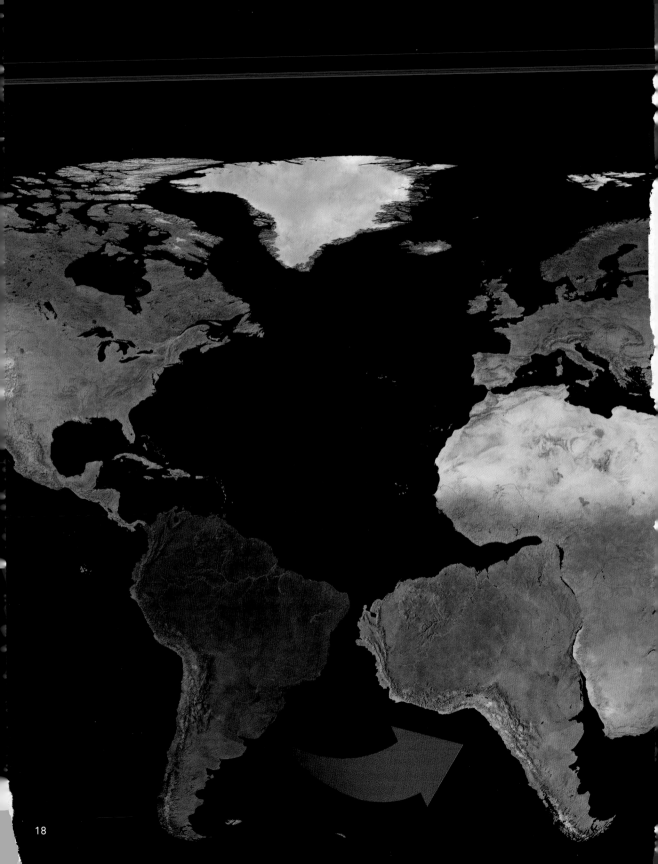

This one has become an icon that is used in many atlases around the world, including by *National Geographic*.

That teacher made the same
mistake that Mark Twain described
in his famous aphorism:

IT AIN'T WHAT YOU DO YOU INTO TROUBLE. I FOR SURE THAT JUST

MARK TWAIN

N'T KNOW THAT GETS
T'S WHAT YOU KNOW
AIN'T SO.

Earth's environment. The most vulnerable part of the Earth's ecological system is the atmosphere. It's vulnerable because it's so thin.

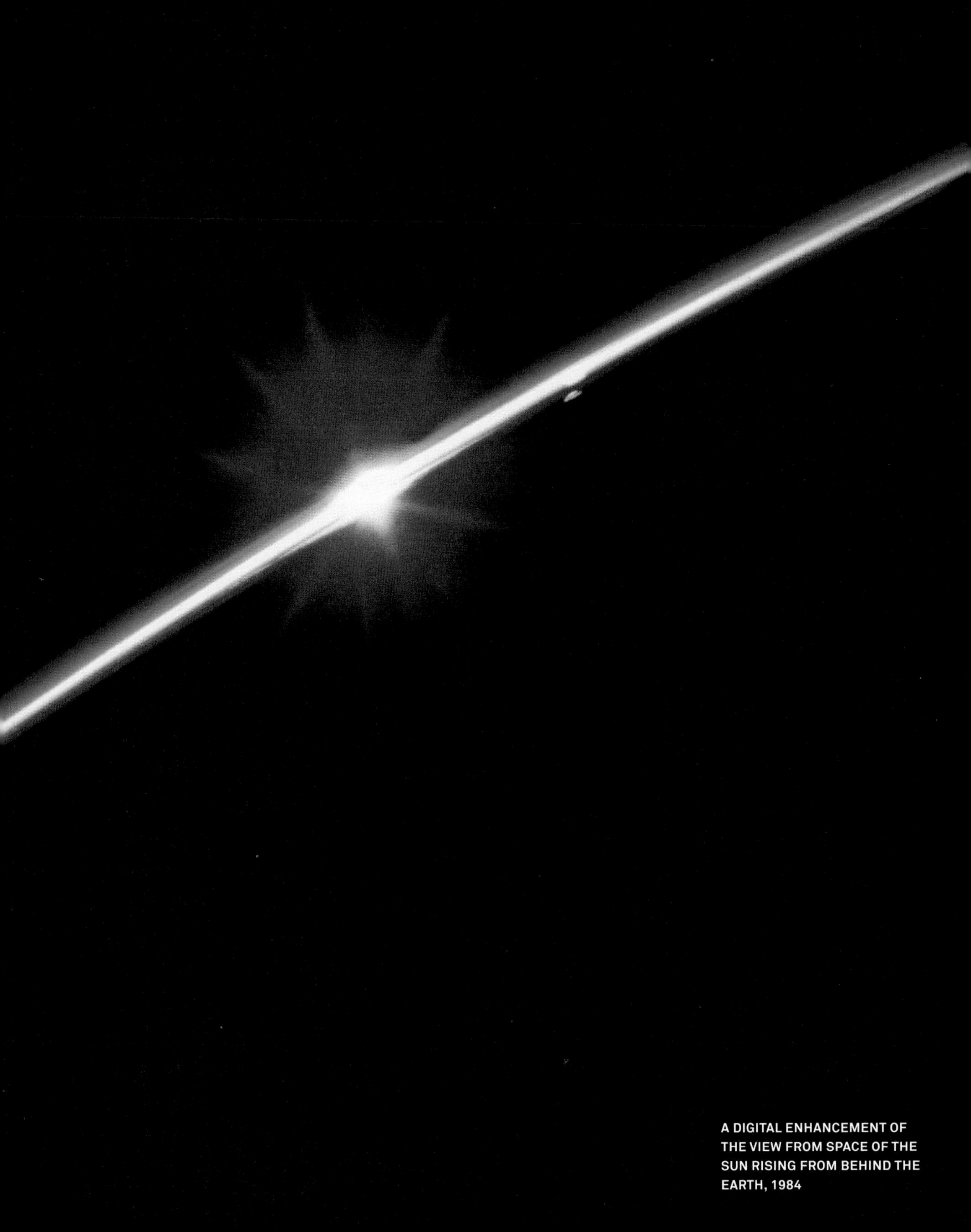

A DIGITAL ENHANCEMENT OF
THE VIEW FROM SPACE OF THE
SUN RISING FROM BEHIND THE
EARTH, 1984

The atmosphere is thin enough that we are capable of changing its composition.

Indeed, the Earth's atmosphere is so thin that we have the capacity to dramatically alter the concentration of some of its basic molecular components. In particular, we have vastly increased the amount of carbon dioxide—the most important of the so-called greenhouse gases.

These images illustrate the basic science of global warming.

The Sun's energy enters the atmosphere in the form of light waves and heats up the Earth. Some of that energy warms the Earth and then is re-radiated back into space in the form of infrared waves.

Under normal conditions, a portion of the outgoing infrared radiation is naturally trapped by the atmosphere—and that is a good thing, because it keeps the temperature on Earth within comfortable bounds. The greenhouse gases on Venus are so thick that its temperatures are far too hot for humans. The greenhouse gases surrounding Mars are almost nonexistent, so the temperature there is far too cold. That's why the Earth is sometimes referred to as the "Goldilocks planet"—the temperatures here have been just right.

The problem we now face is that this thin layer of atmosphere is being thickened by huge quantities of human-caused carbon dioxide and other greenhouse gases. And as it thickens, it traps a lot of the infrared radiation that would otherwise escape the atmosphere and continue out to the universe. As a result, the temperature of the Earth's atmosphere—and oceans— is getting dangerously warmer.

That's what the climate crisis is all about.

WHAT EXACTLY ARE GREENHOUSE GASES?

When we talk about greenhouse gases and climate change, carbon dioxide usually gets the most attention. But there are also some others, although CO_2 is the most important by far.

What all greenhouse gases have in common is that they allow light from the sun to come into the atmosphere, but trap a portion of the outward-bound infrared radiation and warm up the air.

Having some amount of greenhouse gases is beneficial. Without them, the average temperature of the Earth's surface would be right around 0°F—not a very nice place to live. Greenhouse gases also help keep the Earth's surface at a much more hospitable average temperature—almost 59°F. But due to increasing concentrations of human-caused greenhouse gases in modern times, we are raising the planet's average temperature and creating the dangerous changes in climate we all see around us.

CO_2 usually gets top billing in this because it accounts for 80% of total greenhouse gas emissions. When we burn fossil fuels (oil, natural gas, and coal) in our homes, cars, factories, and power plants, or when we cut or burn down forests, or when we produce cement, we release CO_2 into the atmosphere.

Like CO_2, methane and nitrous oxide both predate our presence on the Earth but have gotten huge boosts from us. Sixty percent of the methane currently in the atmosphere is produced by humans; it comes from landfills, livestock farming, fossil-fuel burning, wastewater treatment, and other industry. In large-scale livestock farming, liquid manure is stored in massive tanks that emit methane. Dry manure, left on fields, by contrast, does not. Nitrous oxide (N_2O)—another greenhouse culprit—also occurs naturally, though we have added 17% more of it to the atmosphere just in the course of

our industrial age, from fertilizers, fossil fuels, and the burning of forests and crop residues.

Sulfur hexafluoride (SF_6), PFCs, and HFCs are all greenhouse gases that are produced exclusively by human activity. Not surprisingly, emissions of these gases are on the rise, too. HFCs are used as substitutes for CFCs—which were banned because their emissions in refrigeration systems and elsewhere were destroying the ozone layer. CFCs were also very potent greenhouse gases. PFCs and SF_6 are released into the atmosphere by industrial activities like aluminum smelting and semiconductor manufacturing, as well as the electricity grid that lights up our cities.

And finally, water vapor is a natural greenhouse gas that increases in volume with warmer temperatures, thereby magnifying the impact of all artificial greenhouse gases.

This is the image that first caused me to think about—and then to become intently focused on—global warming. It was shown in the mid-1960s to a small undergraduate class I took taught by the second teacher I want to tell you about: Roger Revelle.

Professor Revelle was the first scientist to propose measuring CO_2 in the Earth's atmosphere. He and the scientist he hired to run the study, Charles David Keeling, began taking daily measurements in the middle of the Pacific Ocean over the big island of Hawaii in 1958.

After the first few years, they had enough data to produce this graphic image, which Professor Revelle showed to my class. It was clear even at this early stage of their experiment that the concentration of CO_2 throughout the Earth's atmosphere was going up at a significant rate.

I asked Revelle why the line marking CO_2 concentration goes up sharply and then down once each year. He explained that the vast majority of the Earth's land mass—as illustrated in this picture— is north of the Equator. Thus, the vast majority of the Earth's vegetation is also north of the Equator.

EQUATOR

I asked Revelle why the line marking CO_2 concentration goes up sharply and then down once each year. He explained that the vast majority of the Earth's land mass—as illustrated in this picture— is north of the Equator. Thus, the vast majority of the Earth's vegetation is also north of the Equator.

EQUATOR

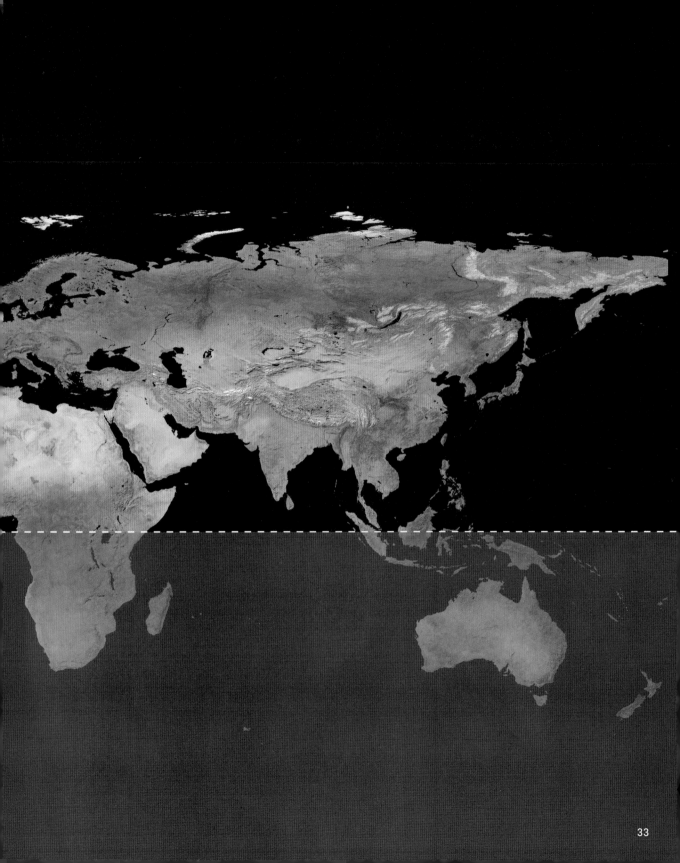

As a result, when the Northern Hemisphere is tilted toward the Sun during the spring and summer, the leaves come out, and as they breathe in CO_2, the amount of CO_2 decreases worldwide.

CO_2 LEVELS

When the Northern Hemisphere is tilted away from the Sun in the fall and winter, the leaves fall, and as they disgorge CO_2, the amount of CO_2 in the atmosphere goes back up again.

It's as if the entire Earth takes a big breath in and out once each year.

CO$_2$ LEVELS

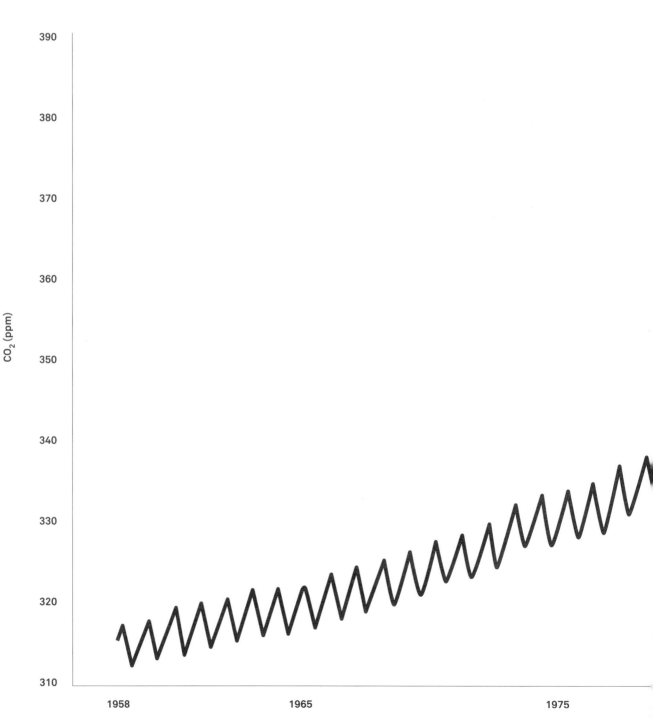

The same pattern of steadily increasing concentrations of CO_2 that was visible after the first several years of Revelle's measurements has continued year by year for almost a half-century. This remarkable and patiently collected daily record now stands as one of the most important series of measurements in the history of science.

The pre-industrial concentration of CO_2 was 280 parts per million. In 2005, that level, measured high above Mauna Loa, was 381 parts per million.

1985   1995   2005

A Scientific Hero

Roger Revelle

◆●◆

As a college undergraduate in the 1960s, I studied under a truly remarkable scientist, Professor Roger Revelle, who was the first person to propose measuring CO_2 in the Earth's atmosphere.

Revelle was an imposing figure who seemed to possess an unusual air of authority and always commanded respect from those who dealt with him. One reason for this may have been that my classmates and I knew that, in addition to being a charismatic teacher, he was also—first and foremost—a hard-nosed scientist devoted to careful and methodical experimentation and patient analysis of the huge amounts of data he collected.

In the 1950s Revelle formed what scientists call a hypothesis, but what seems to me to have been an almost prophetic insight: He saw clearly that the global, post–World War II economic expansion, driven by explosive population growth and fueled mainly by coal and oil, was likely to produce an unprecedented and dangerous increase in the amount of CO_2 in the Earth's atmosphere.

So he proposed and designed a bold new scientific experiment: to collect samples of the CO_2 concentrations high in the Earth's atmosphere from multiple locations every day for many years into the future.

Taking full advantage of the International Geophysical Year due to begin in 1957, Revelle put together funding and hired a young researcher named Charles David Keeling. They established their principal research station at the top of Mauna Loa, the higher of the two massive volcanic mountains on the big island of Hawaii. They chose this spot in the middle of the Pacific Ocean because samples taken there would be untainted by local industrial emissions.

A year later they began launching weather balloons and painstakingly analyzing the amount of CO_2 in the air samples they collected each day. And

Professor Roger Revelle

after the first few years the trend was already clear.

By 1968 when I first walked into his natural sciences classroom, Professor Revelle had become a professor at Harvard, and he shared with my classmates and me the results of the first several years of the CO_2 measurements at Mauna Loa.

I will never forget the graph he drew on the blackboard, nor the dramatic message it conveyed: that something profoundly new was happening to the atmosphere of the entire planet and that this transformation was being caused by human beings.

This was truly startling to me. If we could so quickly significantly change the concentration of such an important atmospheric component—everywhere in the world—then humankind was entering a new relationship with the Earth.

Revelle also expressed concern back then about the unprecedented absorption of CO_2 in the oceans as well as in the atmosphere. He saw early on that the

world's oceans would bear a significant part of the burden of all that extra carbon dioxide released from burning fossil fuels.

Only recently have rigorous new studies confirmed that Revelle was right about that concern too. The world's oceans are becoming more acidic due to the enormous quantities of CO_2 that produce carbonic acid and change the pH of the oceans, first in the colder waters near the poles, but soon—unless we change our ways quickly—throughout the entire ocean.

Even though it was troubling, Professor Revelle's logic all those years ago had the unmistakable ring of truth. It was obvious to our small group that he himself was surprised and disturbed by how quickly the CO_2 was building up. And more important, he understood—and communicated forcefully to us—what the likely implications of this data would turn out to be. He knew that this path our civilization had taken would send us careening toward catastrophe, unless the trend could be reversed.

Clearly, it was difficult for him to tell—and difficult for us to hear—this message that would later become, in our time, an inconvenient truth.

I kept in touch with Professor Revelle after I graduated, and I followed his steady, continuing measurements year after year. When I was elected to Congress to represent Tennessee's fourth congressional district, I helped organize the first congressional hearing on global warming and invited Revelle to be the lead-off witness. I really believed that my congressional colleagues on the committee would experience the same epiphany I had when they heard this great scientist's clear analysis.

I couldn't have been more wrong. The urgency simply didn't come across. This surprised and disappointed me. I'd seriously underestimated the resistance—and disinterest—this alarming prognosis of global warming would meet. It wasn't the last time I would have that experience.

I encountered a similar difficulty when I became a senator and chaired numerous hearings and science roundtables. I ran into it again when some of my colleagues and I failed to pass legislation to cap carbon dioxide emissions. I encountered it in 1987 and 1988 when I first ran for president—in part to gain more attention for this issue—and had great difficulty making it a central focus in the American political dialogue. I experienced it again as vice president when I tried to persuade Congress to pass bold measures to solve the climate crisis, and when I tried to convince the U.S. Senate to ratify the Kyoto Treaty that I helped to write. And I still run into it today.

But I'm not done yet. I am still trying to communicate the implications of the powerful truth first shown to me in that classroom by Roger Revelle.

I am only one of many students inspired by Revelle. Numerous scientists were as well. Most important among them was his research partner, Charles David Keeling, who was honored as a true scientific hero by his colleagues and our nation before his death in 2005. With extraordinary stamina, skill, and precision, Keeling faithfully and painstakingly measured the concentration of CO_2 in our world every day for almost half a century.

Revelle died in 1991, before the world took action on his message. I saw him for the last time in San Diego not long before he died, and I still see his family from time to time. I miss him. He was a truly great man. My life has been changed by his prescient investigation, his wisdom, his dogged clarion call to pay attention to the solid scientific facts, and, perhaps most of all, by his chart.

I still show Revelle's chart of rising CO_2 levels many times each week. It is more elaborate now than when I first saw it—the Mauna Loa measurements now span 48 years. Moreover, with the information gleaned from ice cores drilled in Antarctica and Greenland, the chart has now been extended backward by 650,000 years, and with the help of modern supercomputers and sophisticated climate models, it can be projected many years ahead to measure the future impact of choices we are making today.

It is a tribute to Roger Revelle's scientific brilliance that the essential core of the data we depend on to understand our changing planet is his. And it is a tribute to his wisdom that we are learning about the dangers we face while we still have time to put Earth's balance right again.

TOP: *Revelle testifying at a Congressional hearing, Washington, DC, 1979*
BOTTOM: *Charles Keeling in the lab, La Jolla, CA, 1996*

It is evident in the world around us that very dramatic changes are taking place.

This is Mount Kilimanjaro in 1970 with its fabled snows and glaciers.

MOUNT KILIMANJARO,
TANZANIA, 1970

Here it is just 30 years later—with far
less ice and snow.

MOUNT KILIMANJARO, 2000.

43

A friend of mine, Carl Page, flew over
Mount Kilimanjaro in 2005 and brought
back this picture.

MOUNT KILIMANJARO, 2005

Another friend, Dr. Lonnie Thompson of Ohio State University, is the world's leading expert on mountain glaciers. Here he is at the top of Kilimanjaro in 2000 with the pitiful last remnants of one of its great glaciers.

He predicts that within 10 years there will be no more "Snows of Kilimanjaro."

Our own Glacier National Park will soon need to be renamed "the park formerly known as Glacier."

BOULDER GLACIER, GLACIER
NATIONAL PARK, MT, 1932

Another friend, Dr. Lonnie Thompson of Ohio State University, is the world's leading expert on mountain glaciers. Here he is at the top of Kilimanjaro in 2000 with the pitiful last remnants of one of its great glaciers.

He predicts that within 10 years there will be no more "Snows of Kilimanjaro."

Our own Glacier National Park will soon need to be renamed "the park formerly known as Glacier."

BOULDER GLACIER, GLACIER NATIONAL PARK, MT, 1932

The glacier below, on the left, was a tourist attraction in the 1930s. Now, as seen in the picture on the right, there's nothing there. I climbed to the top of the bigger glacier in this park with one of my daughters in 1997 and heard from the scientists who accompanied us that within 15 years all of the glaciers throughout the park will likely be gone.

BOULDER GLACIER, 1988

Almost all of the mountain glaciers in the world are now melting, many of them quite rapidly. There is a message in this.

PERITO MORENO GLACIER,
PATAGONIA, ARGENTINA, 2003

The red lines show how quickly the
Columbia Glacier in Alaska has receded
since 1980.

WASTAGE OF COLUMBIA
GLACIER, PRINCE WILLIAM
SOUND, AK, 1997

2005

1999

1997

1993

1989

1987

1984

BEFORE 1980

Everywhere in the world the story is the same, including in the Andes in South America.

This is a glacier in Peru just 15 years ago.

QORI KALIS GLACIER, PERU, 1978

This is that same location as it appears in 2006.

QORI KALIS GLACIER, 2006

This beautiful image of a magnificent glacier in Patagonia, on the tip of South America, shows how it stood 75 years ago.

That vast expanse of ice is now gone.

UPSALA GLACIER, PATAGONIA, ARGENTINA, 1928

UPSALA GLACIER, 2004

Throughout the Alps we are witnessing a similar story. Here is an old postcard from Switzerland depicting a scenic glacier early in the last century.

Here is the same site today.

TSCHIERVA GLACIER, SWITZERLAND, 1910

TSCHIERVA GLACIER, 2001

Below is the famous Hotel Belvedere, situated on the Rhone Glacier in Switzerland.

Here is the same site nearly a century later. The hotel is still there—but the glacier is not.

HOTEL BELVEDERE, RHONE GLACIER, SWITZERLAND, 1906

HOTEL BELVEDERE, RHONE GLACIER, 2003

Here is the Roseg Glacier in 1949.

And here it is in 2003.

ROSEG GLACIER, SWITZERLAND, 1949

ROSEG GLACIER, 2003

Here you see a view of the Italian Alps just a century ago.

The same place today looks very different.

ADAMELLO GLACIER, TRENTINO, ITALY, 1880

ADAMELLO GLACIER, 2003

The Himalayan Glaciers on the Tibetan Plateau have been among the most affected by global warming. The Himalayas contain 100 times as much ice as the Alps and provide more than half of the drinking water for 40% of the world's population—through seven Asian river systems that all originate on the same plateau.

Within the next half-century, that 40% of the world's people may well face a very serious drinking water shortage, unless the world acts boldly and quickly to mitigate global warming.

Indus River

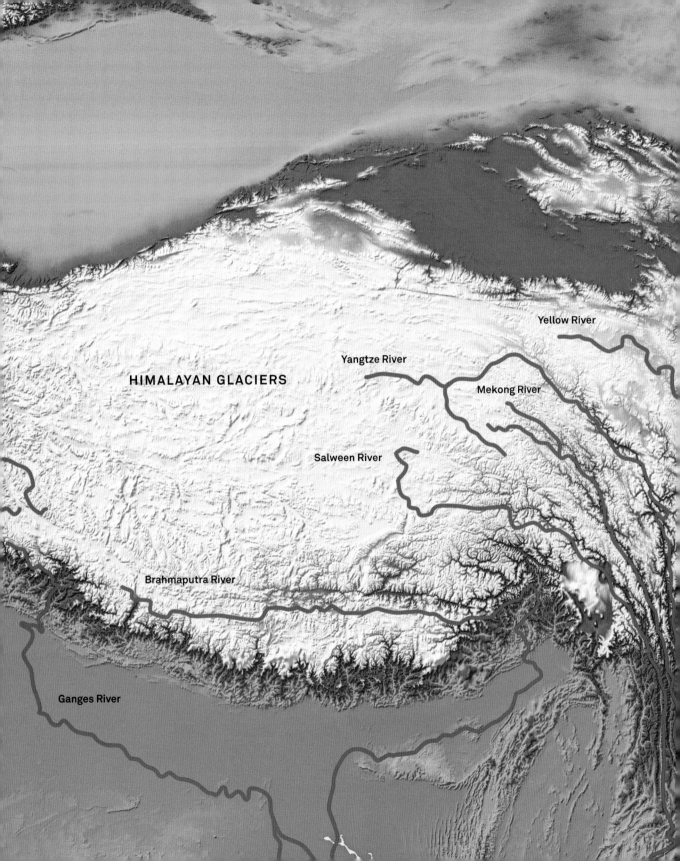

HIMALAYAN GLACIERS

Yellow River

Yangtze River

Mekong River

Salween River

Brahmaputra River

Ganges River

Scientist Lonnie Thompson takes his team to the tops of glaciers all over the world. They dig core drills down into the ice, extracting long cylinders filled with ice that was formed year by year over many centuries.

CAMP OF THOMPSON'S OHIO
STATE UNIVERSITY TEAM, BONA
CHURCHILL COL, AK, 2002

LEFT: THOMPSON'S CREW CORING THROUGH THE ICE, HUASCARÀN, PERU, 1993. RIGHT: RESEARCHER FROM THOMPSON'S TEAM, MOUNT KILIMANJARO, TANZANIA, 2000

Lonnie and his team of experts then examine the tiny bubbles of air trapped in the snow in the year that it fell. They can measure how much CO_2 was in the Earth's atmosphere in the past, year by year. They can also measure the exact temperature of the atmosphere each year by calculating the ratio of different isotopes of oxygen (oxygen-16 and oxygen-18), which provides an ingenious and highly accurate thermometer.

The team can count backward in time year by year—the same way an experienced forester can "read" tree rings— by simply observing the clear line of demarcation that separates each year from the one preceding it, as seen in this unique frozen record.

The thermometer to the right measures temperatures in the Northern Hemisphere over the past 1,000 years.

The blue is cold and the red is hot. The bottom of the graph marks 1,000 years ago and the current era is at the top.

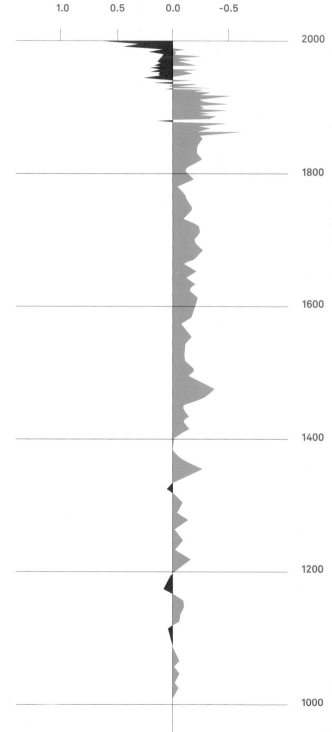

1000 YEARS OF NORTHERN HEMISPHERE
TEMPERATURE (°C)

**ANNUAL LAYERS OF ICE SEEN IN
QUELCCAYA ICE CAP, PERU, 1977**

The correlation between temperature and CO$_2$ concentrations over the last 1,000 years—as measured in the ice core record by Thompson's team—is striking.

Nonetheless, the so-called global-warming skeptics often say that global warming is really an illusion reflecting nature's cyclical fluctuations. To support their view, they frequently refer to the Medieval Warm Period.

But as Dr. Thompson's thermometer shows, the vaunted Medieval Warm Period (the third little red blip from the left, below) was tiny compared to the enormous increases in temperature of the last half-century (the red peaks at the far right of the chart).

1000 YEARS OF NORTHERN HEMISPHERE TEMPERATURE (°C)

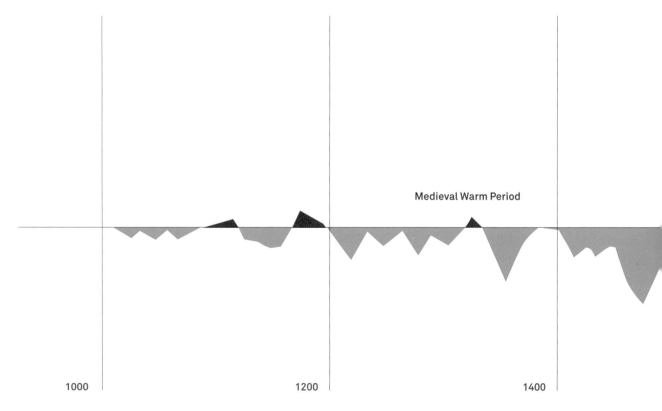

Medieval Warm Period

1000   1200   1400

Year

Those global-warming skeptics—a group diminishing almost as rapidly as the mountain glaciers—launched a fierce attack against another measurement of the 1,000-year correlation between CO_2 and temperature known as "the hockey stick," a graphic image representing the research of climate scientist Michael Mann and his colleagues. But in fact, scientists have confirmed the same basic conclusions in multiple ways—with Thompson's ice core record as one of the most definitive.

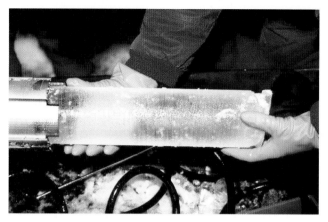

GLACIOLOGIST REMOVING AN ICE CORE, ANTARCTICA, 1993

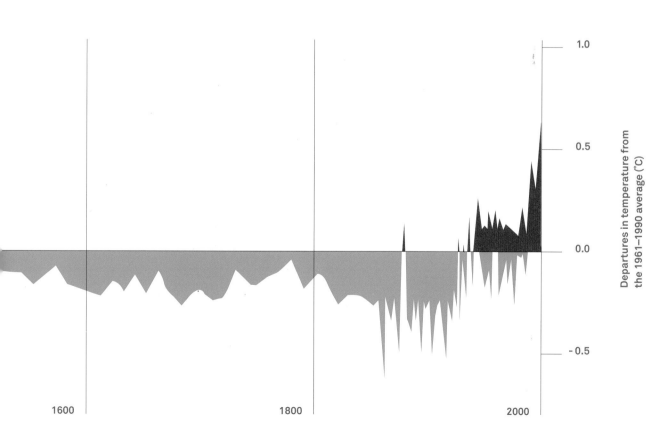

Departures in temperature from the 1961–1990 average (°C)

1.0

0.5

0.0

- 0.5

1600

1800

2000

SOURCE: IPCC

In Antarctica, measurements of CO_2 concentrations and temperatures go back 650,000 years.

The blue line below charts CO_2 concentrations over this period.

The top right side of the blue line represents the present era, and that first dip down as you move from right to left is the last ice age. Then, continuing to the left, you can see the second most recent ice age, the third and fourth most recent ice ages, and so on. In between them are periods of warming.

At no point in the last 650,000 years before the preindustrial era did the CO_2 concentration go above 300 parts per million.

The gray line shows the world's temperature over the same 650,000 years.

Here is an important point. If my classmate from the sixth grade were to see this—you remember, the guy who asked about South America and Africa—he would ask, "Did they ever fit together?"

The answer from scientists would be, "Yes, they do fit together."

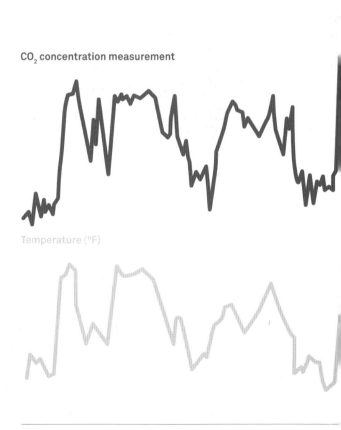

CO₂ concentration measurement

Temperature (°F)

600,000   500,000

SOURCE: *SCIENCE* MAGAZINE

It's a complicated relationship, but the most important part of it is this: When there is more CO_2 in the atmosphere, the temperature increases because more heat from the Sun is trapped inside.

Here's where CO_2 is now—way above anything measured in the prior 650,000-year record.

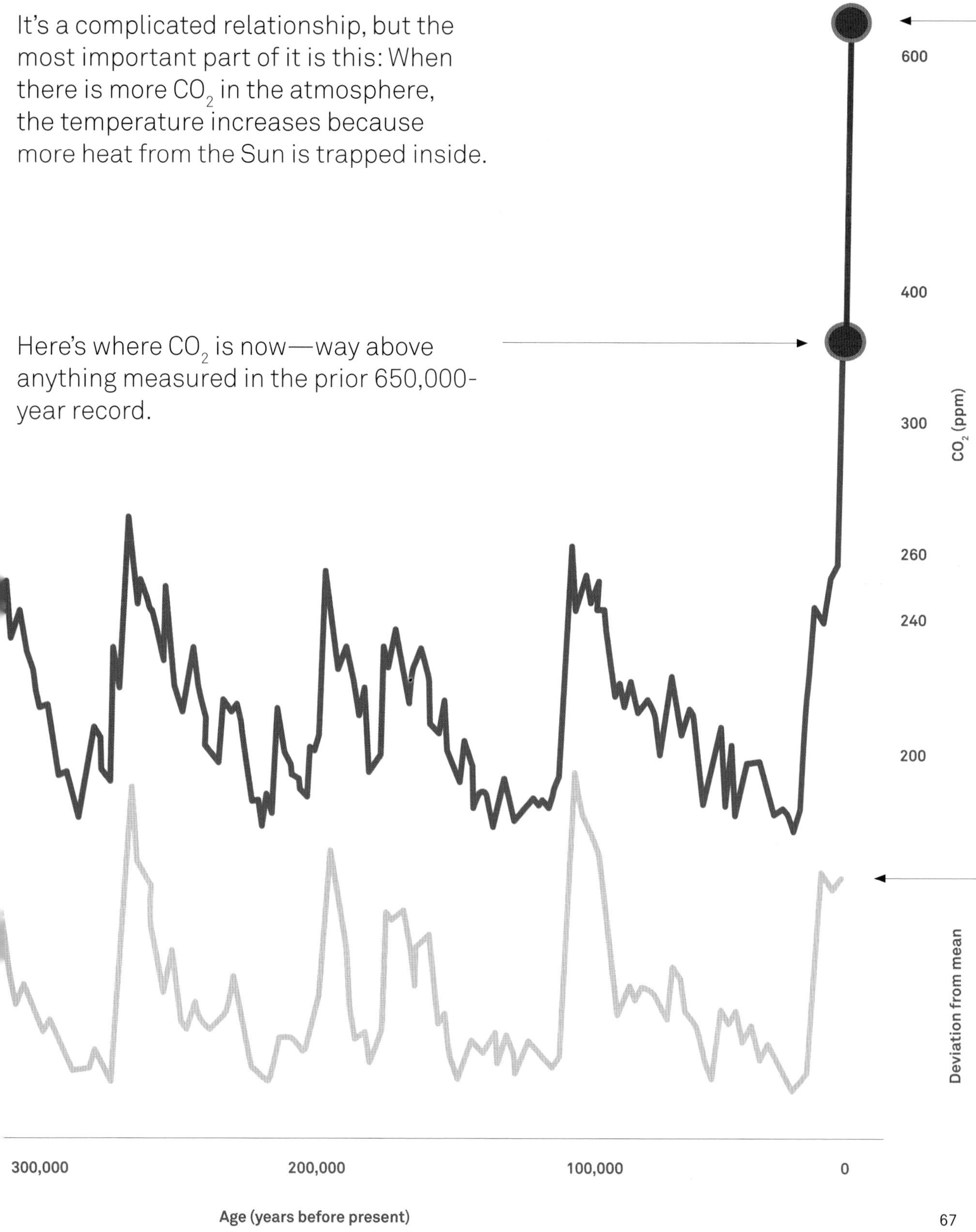

And within 45 years, this is where the CO_2 equivalent levels will be if we do not make dramatic changes quickly.

There is not a single part of this graph—no fact, date, or number—that is controversial in any way or in dispute by anybody.

To the extent that there is a controversy at all, it is that a few people in some of the less responsible coal, oil, and utility companies say, "So what? That's not going to cause any problem."

But if we allow this to happen, it would be deeply and unforgivably immoral. It would condemn coming generations to a catastrophically diminished future.

The top right point of this gray line shows current global temperatures. And the bottom right point marks the depth of the last ice age. That short distance—about an inch in the graph—represents the difference, in Chicago, between a nice day and a mile of ice over your head. Imagine what three times that much on the warm side would mean.

CO_2 (ppm)

600

400

300

260

240

200

Deviation from mean

300,000   200,000   100,000   0

Age (years before present)

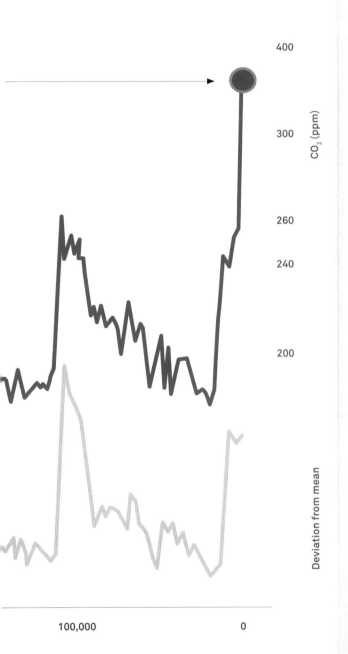

600

400

300

260

240

200

CO$_2$ (ppm)

Deviation from mean

100,000   0

A Turning Point

**I was handed not just a second chance,
but an obligation to pay attention
to what matters**

Some events stay with you always, and change the way you look at everything, no matter how many years go by. My son's serious accident when he was very young was that kind of event for me. It turned my life upside down and shook it until everything fell out. It was every parent's nightmare, and I will never forget any part of it.

It was a bright spring day in early April 1989. Tipper and I had taken our son to the season-opening Orioles game in Baltimore. We'd had a wonderful time, and as we left the stadium, my son's hand was in mine. Albert was just six and already loved baseball.

We stopped at the curb with some neighbors after a long walk toward the lot where our car was parked. Suddenly one of Albert's friends, who was walking just in front of us with his dad, bolted off the sidewalk and took off running as fast as he could across the busy one-way street—even though traffic was heading toward him from only half a block away.

Then, in the next instant, with no warning, my son pulled his hand out of mine and jumped off the curb, running to

chase his friend across the street. But just before he got to the far lane, he was hit by a speeding car. I watched what no parent should ever see: My son was knocked into the air with a horrible thud and then hit the pavement 30 feet away from the point of impact, scraping the pavement until he came to rest, motionless and silent.

I don't know how many times I've relived those horrifying seconds, watching my precious child hurtle up and out of reach, as I squeezed my fist tightly in a futile effort to hold the little hand that was already gone from mine.

I have come to believe that we were literally in the company of angels that day. Two off-duty nurses from Johns Hopkins Hospital who had also gone to the game had brought their bags of emergency medical gear with them—just in case—and as I knelt by my son and prayed aloud, they appeared by his side and skillfully tended to him till the ambulance arrived.

Waiting to hear that rescue siren was the most excruciating six minutes of my life. Tipper and I both knelt next to Albert, holding onto him, talking to him,

and praying. I have never felt so desperate or so helpless.

Albert's life was saved by those nurses on the street and by the doctors and nurses at nearby Johns Hopkins Hospital, where he was quickly taken by the ambulance. He had suffered a concussion, a broken collarbone, broken ribs, a compound fracture of his thighbone, and massive internal injuries, including a ruptured spleen (most of which had to be removed the following day), as well as a bruised lung and pancreas, and a fractured kidney. He had second-degree

burns where he'd skidded on the concrete and damage to the large cluster of nerves that ran down from the spinal cord through the shoulder to his right arm, causing him to lose the use of his right arm entirely for almost a year.

Tipper and I lived at the hospital for a month, and finally—blessedly—Albert came home in a full-body cast. After months of rehabilitation, with our three daughters helping around the clock (even taking turns relieving us in the middle of the night when Albert had to be manually turned in his bed), he

recovered. And within a year, he was completely healed and back to full strength in every way.

I tell this story because it was a turning point that changed me in ways I couldn't have imagined. And even though it is hard to describe in words the linkage between the searing pain of this event and the new outlook I formed on what is really important, inside me it is there—always. Suddenly, the events that packed my schedule—once so seemingly urgent—were revealed as truly insignificant. I realized how trivial those events were that a month earlier had seemed so weighty and began looking at my whole life through the same new lens. I asked myself how did I really want to spend my time on Earth? What really matters?

For me, the first answer was my family: my wife and my children. I made immediate changes to prioritize time with them—each one of them individually and all of us as a family together—in a way I had not done before. I subordinated everything else in my schedule and weekly routine to put time with them first. Quality time and ample time.

I also reevaluated the nature of my public service. I questioned what it really means to "serve." The environment had, for years, been at the forefront of my policy concerns, but it had been competing for attention with a lot of other issues. Now, in this comprehensive and soul-searching rethinking of how I would spend my time, the global environment trumped all other concerns. I realized this was the crisis that loomed largest and should occupy the bulk of my efforts and ingenuity.

It was during Albert's recovery that I started writing my first book, *Earth in the Balance*. It was then that I started putting together the first version of the slide show. Not only was it a way to warn my

LEFT: *Al and Albert, Carthage, TN, 1990;*
BOTTOM LEFT: *Shaving with Albert, 1984*

TOP RIGHT: *Al at Albert's bedside, Johns Hopkins Hospital, Baltimore, MD, 1989;* MIDDLE: *Albert and sister Karenna at home in Virginia, 1989;* BOTTOM RIGHT: *Al and Albert after climbing to the top of Mount Rainier, WA, 1999*

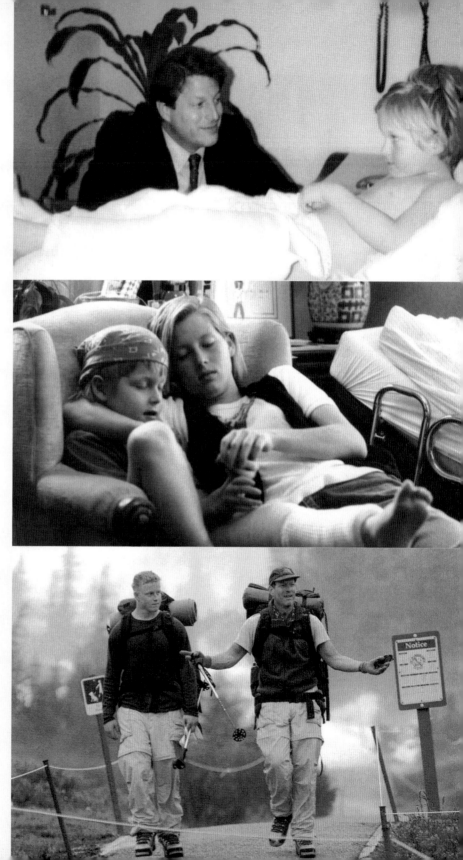

fellow Americans of the spiraling disaster to which we are all—wittingly or not—contributing, it was also a way to put my own priorities in order.

I would give anything to redo that awful day long ago, to glue my young boy's hand to my own, to miss that Orioles game entirely. But I know I can't. And I feel so grateful for the healing and grace that was given to him and to our family. A child acts on impulse, and suddenly his parent cannot protect him. Sometimes the worst does happen. We were lucky enough to get a reprieve that afternoon. We have been blessed with four healthy, wonderful, thriving grown children. And now we dote on our own grandchildren.

But I truly believe I was handed not just a second chance, but an obligation to pay attention to what matters and to do my part to protect and safeguard it, and to do whatever I can at this moment of danger to try to make sure that what is most precious about God's beautiful Earth—its liveability for us, our children, future generations—doesn't slip from our hands.

This graph charts the actual measurements of global temperature increases since the Civil War. In any given year it might seem as if the average global temperature is going down, but the overall trend is very clear. And in recent years the rate of increase has been accelerating.

In fact, if you look at the 21 hottest years measured, 20 of the 21 have occurred within the last 25 years.

GLOBAL TEMPERATURE SINCE 1860:
COMBINED ANNUAL LAND, AIR, AND SEA SURFACE
TEMPERATURES FROM 1860 TO 2005

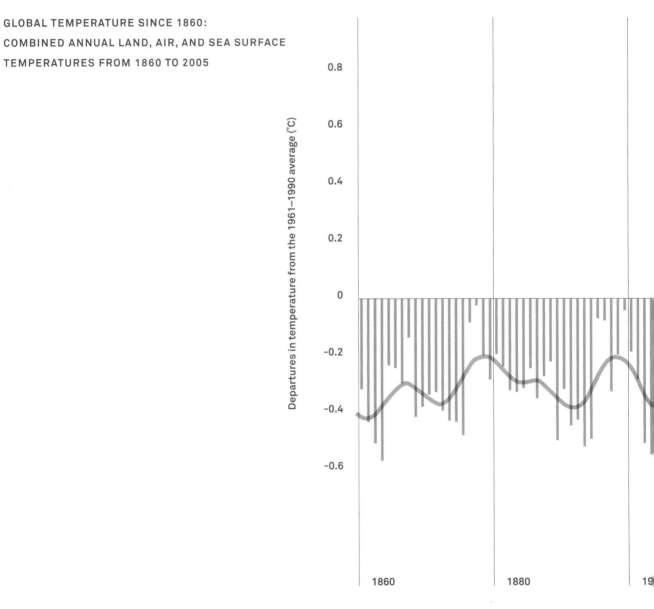

The hottest year recorded during this entire period was 2005.

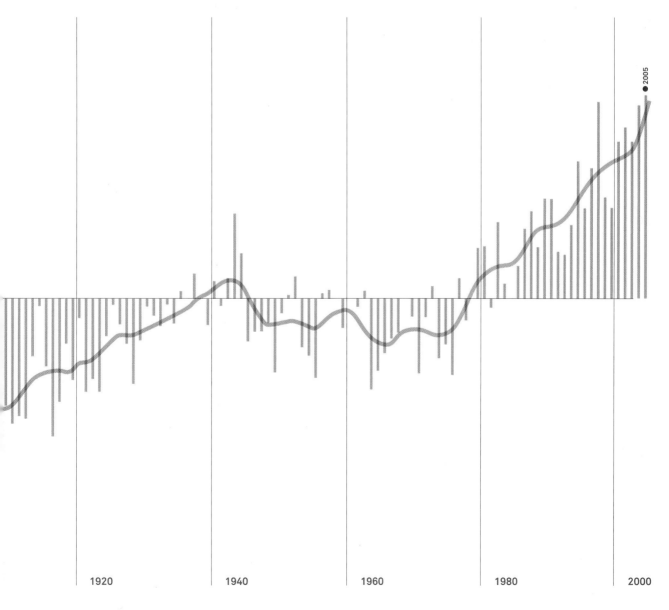

2005

1920   1940   1960   1980   2000

SOURCE: IPCC

We have already begun to see the kind of heatwaves that scientists say will become much more common if global warming is not addressed. In the summer of 2003 Europe was hit by a massive heatwave that killed 35,000 people.

MUNICH ZOO DURING HEAT WAVE, MUNICH, GERMANY, 2003

In the summer of 2005 many cities in the American West broke all-time records for high temperatures and for the number of consecutive days with temperatures of 100°F or more.

In all, more than 200 cities and towns in the West set all-time records.

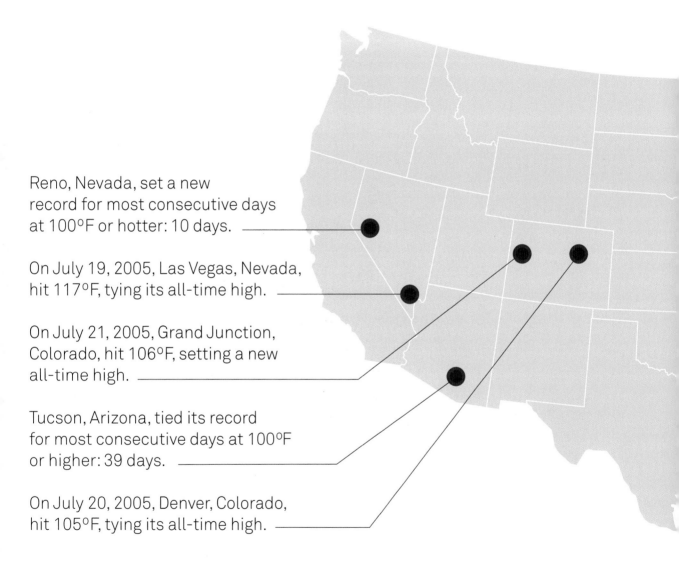

Reno, Nevada, set a new record for most consecutive days at 100°F or hotter: 10 days.

On July 19, 2005, Las Vegas, Nevada, hit 117°F, tying its all-time high.

On July 21, 2005, Grand Junction, Colorado, hit 106°F, setting a new all-time high.

Tucson, Arizona, tied its record for most consecutive days at 100°F or higher: 39 days.

On July 20, 2005, Denver, Colorado, hit 105°F, tying its all-time high.

And in the East, a number of cities set daily temperature records, including, significantly, New Orleans.

On June 23, 2005, La Crosse, Wisconsin, hit 98°F, a new daily high.

On July 27, 2005, Newark, New Jersey, hit 101°F, a new high for that date.

On July 26, 2005, Raleigh-Durham, North Carolina, hit 101°F.

On July 26, 2005, Florence, South Carolina, hit 101°F.

On July 25, 2005, the Louis Armstrong Airport in New Orleans, Louisiana, hit 98°F.

These temperature increases are taking place all over the world, including in our oceans.

Many people say about the rising temperatures, "Oh, it's just natural variability. These things go up and down, so we shouldn't worry."

And indeed there is always a lot of variability in temperatures, and that's true in the oceans as well. The blue line on the graph below shows the normal range of temperature variability in the world's oceans over the past 60 years.

But scientists who specialize in global warming have been using evermore accurate computer models that long ago predicted a much higher range of ocean temperatures as a result of man-made global warming. What the computers told us would happen as a result of climate change is shown in the thick green part of the graph below, which began to diverge from the boundaries of natural variability in the mid-1970s.

What have the *actual* ocean temperatures been?

PREDICTED AND OBSERVED UPPER-LEVEL OCEAN TEMPERATURES, 1940–2004

□ Predicted natural variability

▦ Expected variation due to human causes

■ Actual observed temperatures

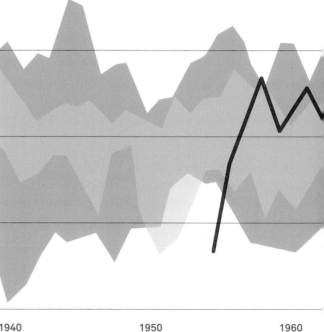

| 1940 | 1950 | 1960 |

The red line on the graph is new. It shows actual ocean temperatures, which have been painstakingly compiled from measurements of all the world's oceans taken over the past 60 years.

The actual ocean temperatures are completely consistent with what has been predicted as a result of man-made global warming. And they're way above the range of natural variability.

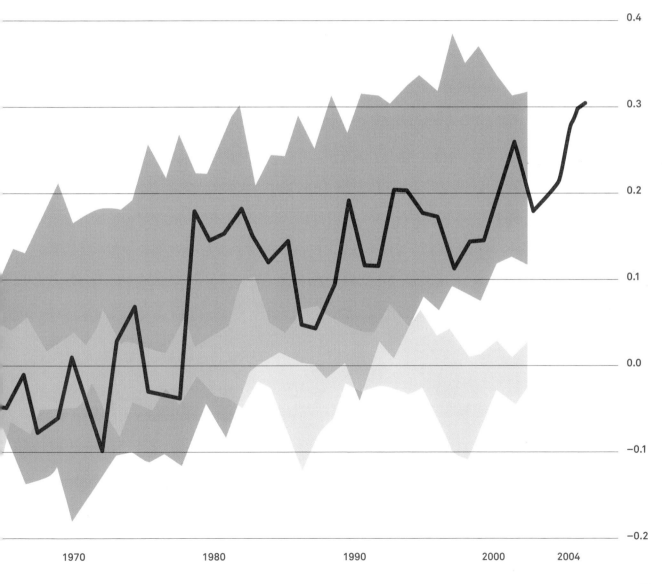

As the oceans get warmer, storms get stronger. In 2004, Florida was hit by four unusually powerful hurricanes.

HURRICANE JEANNE, FL,
SEPTEMBER 2004

A growing number of new scientific studies are confirming that warmer water in the top layer of the ocean can drive more convection energy to fuel more powerful hurricanes.

There is less agreement among scientists about the relationship between the total number of hurricanes each year and global warming—because a multi-decade natural pattern has a powerful influence on hurricane frequency. But there is now a strong, new emerging consensus that global warming is indeed linked to a significant increase in both the duration and intensity of hurricanes.

Brand-new evidence is causing some scientists to assert that global warming is even leading to an increased frequency of hurricanes, overwhelming the variability in frequency long understood to be part of natural deep-current cycles.

As the United States was being hit by numerous large hurricanes in 2004, Japan's weather did not get as much attention in the Western media.

Yet that same year, Japan set an all-time record for typhoons. The previous record was seven. In 2004, 10 typhoons hit Japan. Typhoons, hurricanes, and cyclones are all the same weather phenomena, depending on the ocean in which they originate. In the spring of 2006, Australia was hit by several unusually strong, Category 5 cyclones, including Cyclone Monica, the strongest cyclone ever measured, off the coast of Australia—stronger than Hurricanes Katrina, Rita, or Wilma.

TYPHOON NAMTHEUN, OFF
THE COAST OF JAPAN, JULY 2004

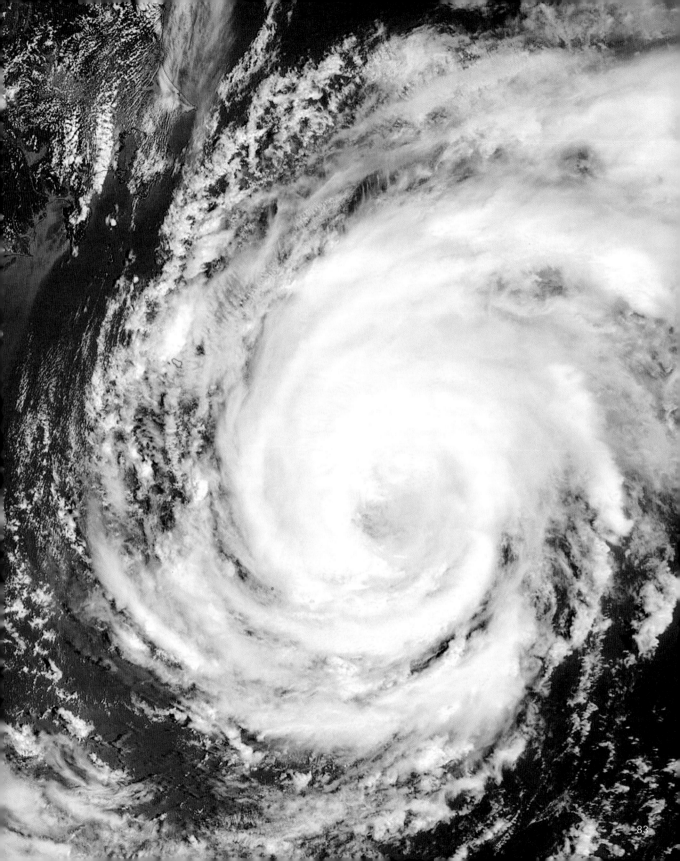

The science textbooks had to be rewritten in 2004. They used to say, "It's impossible to have hurricanes in the South Atlantic." But that year, for the first time ever, a hurricane hit Brazil.

HURRICANE CATARINA, BRAZIL, MARCH 2004

Also in 2004, the all-time record for tornadoes in the United States was broken.

Hard on the heels of 2004 came the record-breaking summer of 2005. Several hurricanes hit the Caribbean and the Gulf of Mexico early in the season, including Hurricane Dennis and Hurricane Emily, which caused significant damage.

DAMAGE CAUSED BY HURRICANE EMILY, LA PESCA, MEXICO, JULY 2005

The emerging consensus linking global warming to the increasingly destructive power of hurricanes has been based in part on research showing a significant increase in the number of category 4 and 5 hurricanes.

A separate study predicts that global warming will increase the strength of the average hurricane a full half-step on the well known five-step scale.

The National Oceanic and Atmospheric Administration summarized some of the basic elements common to these new research studies in the graph shown below.

As water temperatures go up, wind velocity goes up, and so does storm moisture condensation.

HURRICANE INTENSITY GROWS AS OCEANS HEAT UP

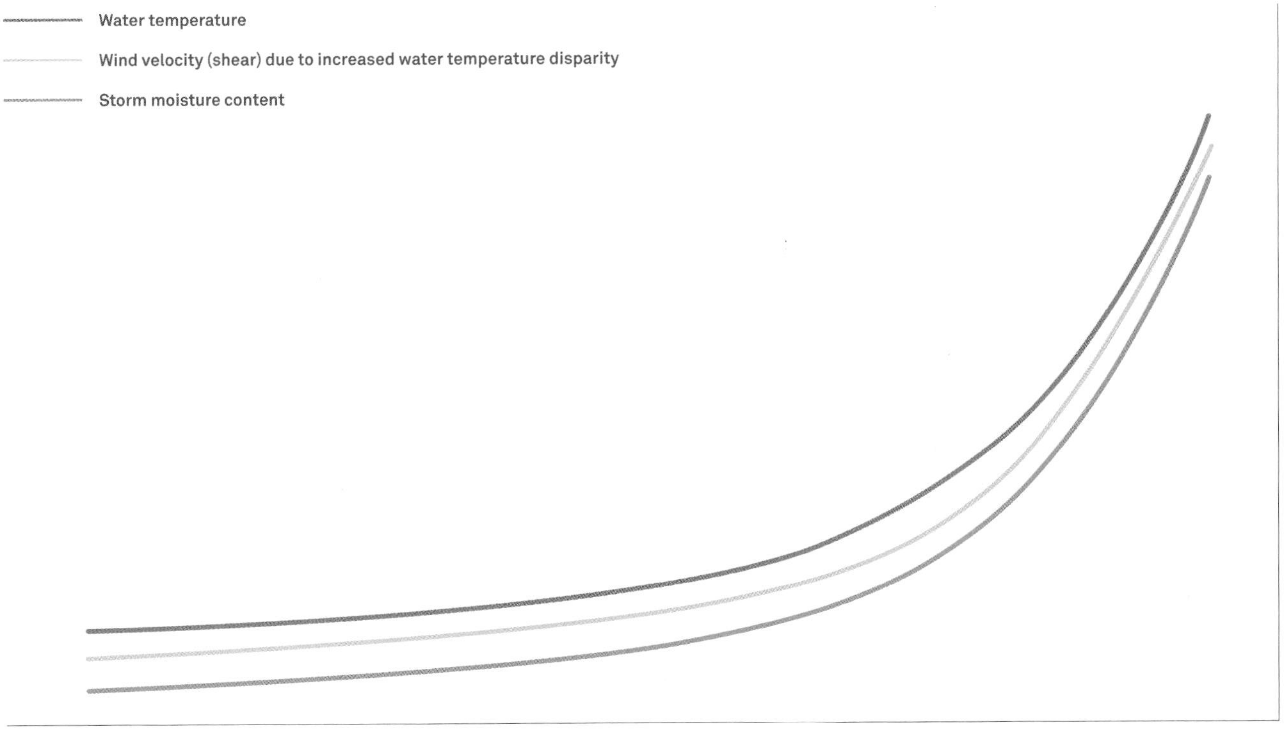

———— Water temperature

———— Wind velocity (shear) due to increased water temperature disparity

———— Storm moisture content

Below is the largest oil platform in the world, BP's Thunder Horse platform, 150 miles southeast of New Orleans, after Hurricane Dennis hit the Gulf on July 11, 2005. In April 2006, one-third of the Gulf's oil-producing facilities, including Thunder Horse, remained crippled.

DAMAGED THUNDER HORSE OIL PLATFORM, OFF THE GULF COAST, LA, JULY 2005

The 13,000-ton oil platform below was driven into this bridge in Mobile, Alabama, later in the 2005 hurricane season.

**OIL RIG WEDGED UNDER THE
COCHRANE BRIDGE, MOBILE, AL,
AUGUST 2005**

On July 31, 2005, less than a month before Hurricane Katrina hit the United States, a major study from MIT supported the scientific consensus that global warming is making hurricanes more powerful and more destructive.

MAJOR STORMS SPIN ATLANTIC AND THE PA 1970S HAVE INCREAS AND INTENSITY BY AB

MIT STUDY, 2005

NING IN BOTH THE
CIFIC SINCE THE
ED IN DURATION
OUT 50 PERCENT.

And then came Katrina. When it first hit Florida on its way into the Gulf on the morning of August 26, 2005, it was only a category 1 storm, but it killed a dozen people and caused billions of dollars in damage.

Then, it passed over the unusually warm waters of the Gulf of Mexico. By the time Katrina hit New Orleans, it was a massive and powerfully destructive storm.

TIME-LAPSE SATELLITE IMAGES OF HURRICANE KATRINA OVER SOUTHERN UNITED STATES, SEPTEMBER 2005

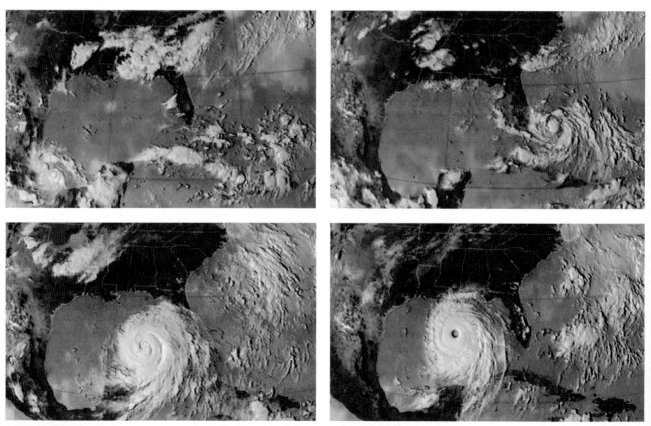

The consequences were horrendous. There are no words to describe them.

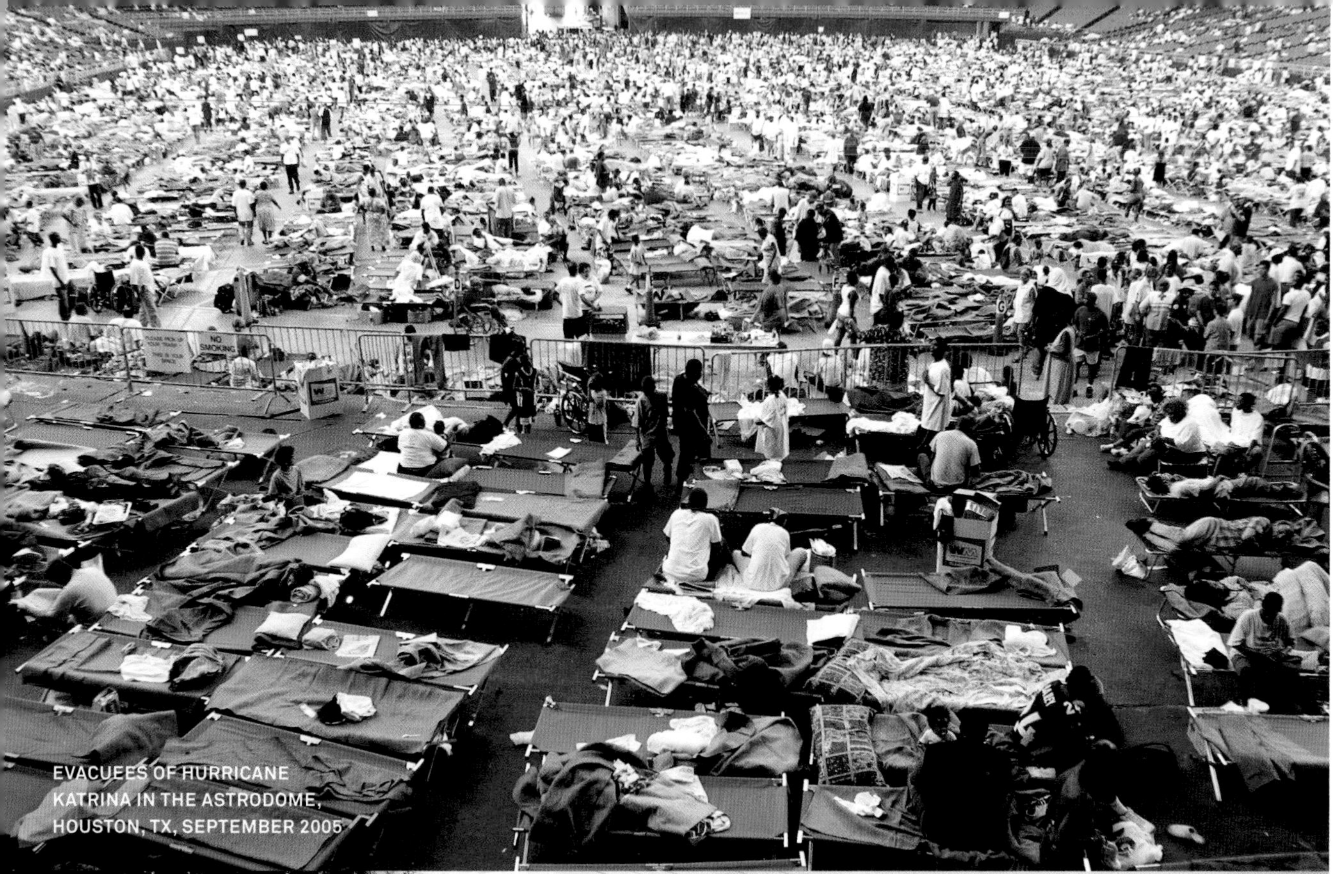

EVACUEES OF HURRICANE
KATRINA IN THE ASTRODOME,
HOUSTON, TX, SEPTEMBER 2005

LOWER NINTH WARD, NEW
ORLEANS, LA, FEBRUARY 2006

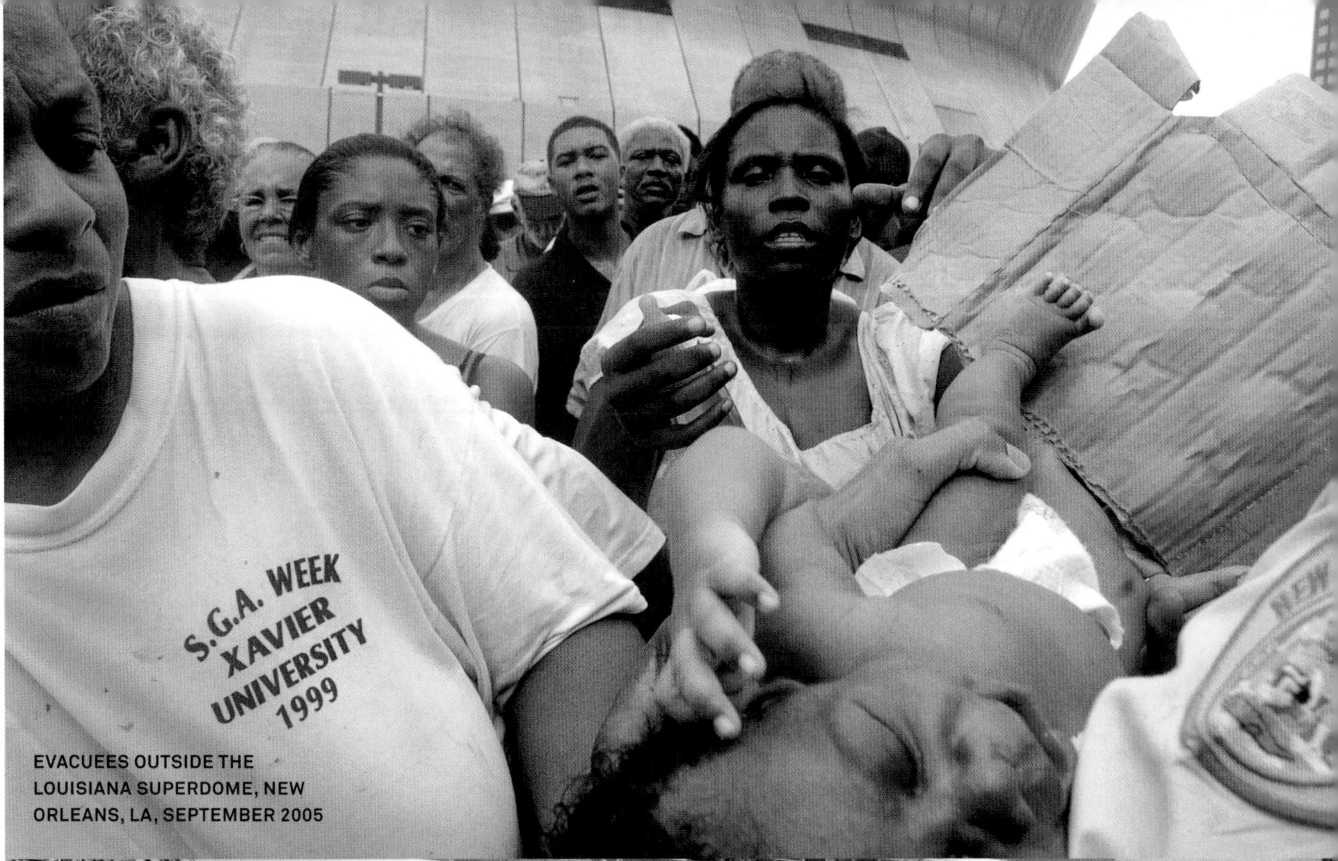

EVACUEES OUTSIDE THE
LOUISIANA SUPERDOME, NEW
ORLEANS, LA, SEPTEMBER 2005

DEVASTATION OUTSIDE THE
SUPERDOME, NEW ORLEANS,
LA, SEPTEMBER 2005

NEW ORLEANS, LA,
SEPTEMBER 2005

THE ERA OF PROCRAS
MEASURES, OF SOOTH
EXPEDIENTS, OF DELA
CLOSE. IN ITS PLACE W
PERIOD OF CONSEQUE

TINATION, OF HALF-
ING AND BAFFLING
YS, IS COMING TO ITS
E ARE ENTERING A
NCES.

The insurance industry is one business sector that's already feeling the unmistakable economic impact of global warming. Over the last three decades, insurance companies have seen a 15-fold increase in the amount of money paid to victims of extreme weather. Hurricanes, floods, drought, tornadoes, wildfires, and other natural disasters have caused these losses. Many can be linked to factors that are worsened by global warming. These natural disasters can be economically—as well as personally—devastating. Hurricane Katrina alone caused an estimated $60 billion in insured losses.

Some companies in the risk management business have recognized the trend and recently convened a task force to analyze the potential impact of climate change on their businesses. More is at stake than the future financial health of the industry and the affordability of premiums for most Americans. The ripple effect will very likely extend far beyond insurance ledgers. Many pensions and mutual funds have insurance companies in their portfolios, and could also be affected.

Insurers base their rates—the amount you pay to protect your home against disaster—on their ability to calculate the risk of unexpected events. When extreme weather stops following predictable, historical patterns—as it appears is already happening—companies can no longer estimate risk accurately, which in turn makes it difficult to project what their losses will be. The only way to stay

in business under these conditions would be to raise premiums for all insurance holders, or to stop offering insurance in particularly risky areas, such as Florida and the Gulf coast, which already face increasingly devastating weather every summer.

As one business leader put it, insurance companies face "a perfect storm of rising weather losses, rising global temperatures, and more Americans than ever living in harm's way."

GREAT WEATHER AND FLOOD CATASTROPHES: LOSSES IN BILLIONS OF U.S. DOLLARS

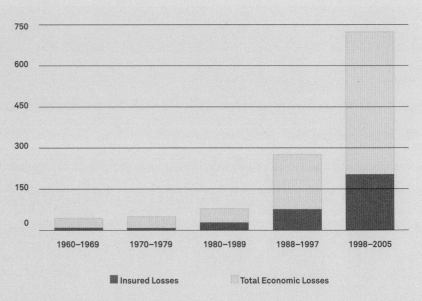

SOURCE: MUNICH RE, SWISS RE, 2005, SIGMA FIGURES AS OF 12/20/05

Just three weeks after Katrina, another hurricane that reached category 5 strength—Rita—hit the U.S. coast not too far west of Katrina's landfall. This time, it hit less populated areas, though the damage and suffering it caused were still devastating.

Then, just a few weeks after Rita, Hurricane Wilma became—while still out at sea—the strongest hurricane ever measured.

It traveled back eastward from Mexico's Yucatan Peninsula to southern Florida, causing massive damage and leaving thousands without water or electricity for weeks.

And before Wilma left the scene, something new happened: We ran out of names. For the first time in history, the World Meteorological Organization had to start using the letters of the Greek alphabet to name the hurricanes and tropical storms that continued on into December—well past the end of the 2005 hurricane season.

HURRICANE RITA AFTERMATH, CAMERON, LA, SEPTEMBER 2005

Here are all 27
of them.

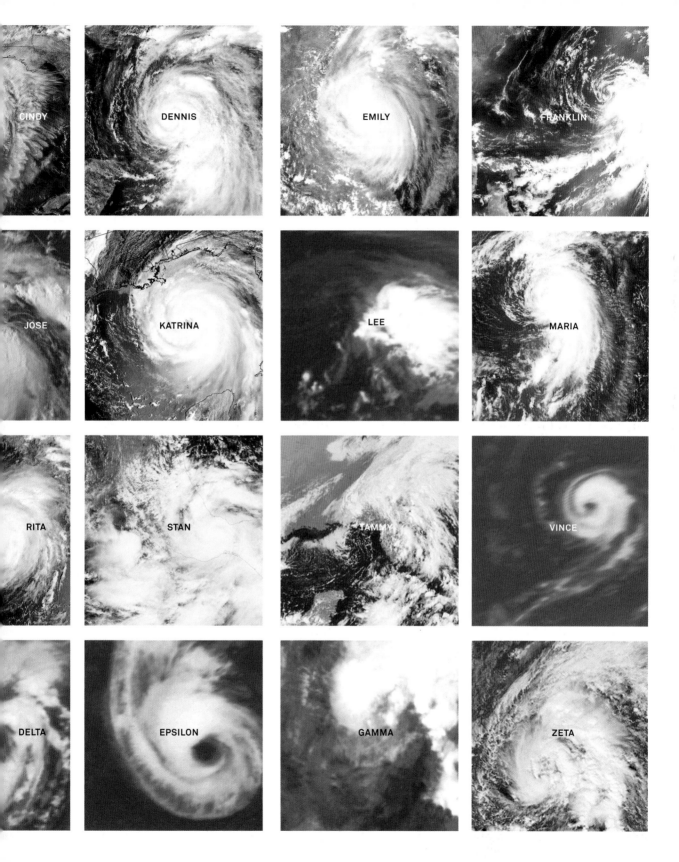

Warmer water increases the moisture content of storms, and warmer air holds more moisture. When storm conditions trigger a downpour, more of it falls in the form of big, one-time rainfalls and snow-falls. Partly as a result, the number of large flood events has increased decade by decade, on every continent.

In many areas of the world, global warming also increases the percentage of annual precipitation that falls as rain instead of snow, which has led to more flooding in spring and early summer.

In 2005 Europe had a year of unusual catastrophes very similar to the one in the United States.

NUMBER OF MAJOR FLOOD EVENTS BY CONTINENT AND DECADE

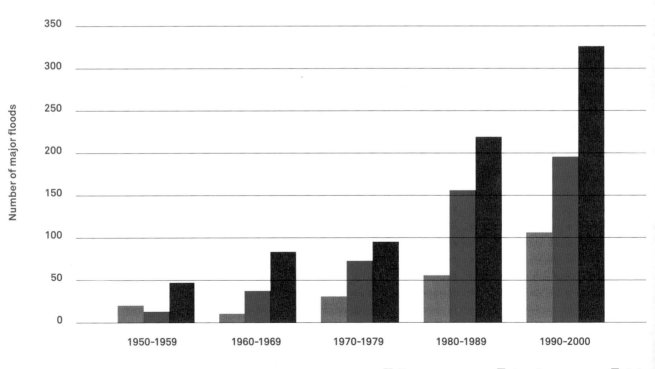

SOURCE: MILLENNIUM ECOSYSTEM ASSESSMENT

Europe   Americas   Asia

While the United States was ending a seemingly unprecedented string of large hurricanes in 2005, Europe was experiencing a disastrous number of floods. United Press International summarized the feelings of many Europeans on August 26, 2005, when it reported: "Nature is going crazy in Europe."

FLOOD DAMAGE, BRIENZ, SWITZERLAND, AUGUST 2005

FLOODED SCHWEIZERHOF
QUAY, LUCERNE, SWITZERLAND,
AUGUST 2005

It was almost like a nature hike through the book of Revelation.

Flooding in Asia has also increased dramatically. In July 2005, Mumbai, India, received 37 inches of rain in 24 hours. It was, by far, the largest downpour that any city in India has ever received in one day. Water levels reached seven feet. The death toll in western India reached 1,000. This photograph shows rush hour the next day.

**COMMUTERS AFTER
TORRENTIAL RAINS, MUMBAI,
INDIA, JULY 2005**

There has also been record flooding in China, which, as one of the planet's oldest civilizations, keeps the best flood records of any nation in the world.

Recently, for example, there were huge floods in the Sichuan and Shandong provinces. Paradoxically, however, global warming causes not only more flooding, but also more drought. The nearby Anhui province was continuing to suffer a severe drought at the same time the neighboring areas were flooding.

One of the reasons for this paradox has to do with the fact that global warming not only increases precipitation world-wide but at the same time causes some of it to relocate.

FLOODING IN SHANGDONG
PROVINCE, CHINA, JUNE 2005

DROUGHT IN ANHUI PROVINCE,
CHINA, JUNE 2005

This graphic shows that, overall, the amount of precipitation has increased globally in the last century by almost 20%.

However, the effects of climate change on precipitation are not uniform. Precipitation in the 20th century increased overall, as expected with global warming, but in some regions precipitation actually decreased.

Decreased precipitation
Increased precipitation

-50%   -40%   -30%   -20%   -10%

The blue dots mark the areas with increased precipitation—the larger the dot, the larger the increase. The orange dots show the places and amounts of decreased precipitation.

Sometimes the effects of such a large shift can be devastating. For example, focus on the part of Africa just on the edge of the Sahara.

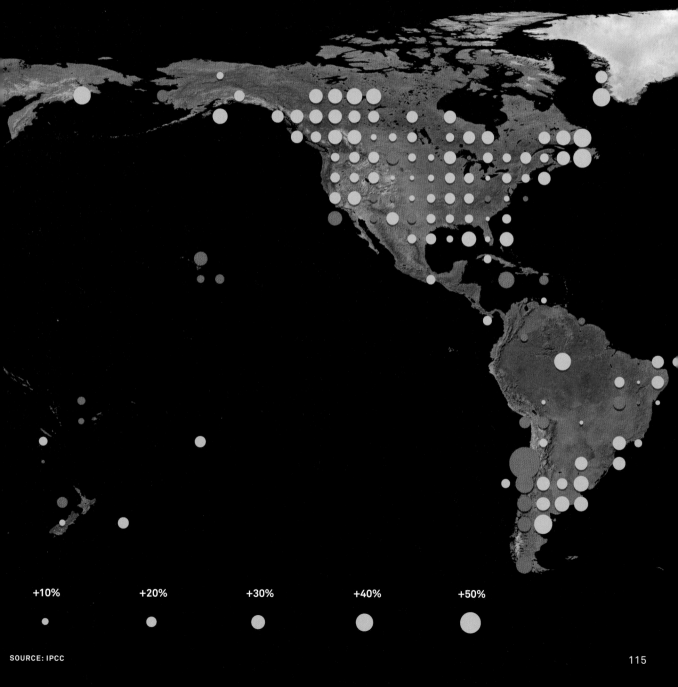

+10%   +20%   +30%   +40%   +50%

SOURCE: IPCC

Unbelievable tragedies have been unfolding in the part of Africa that includes southern Sudan to the east of Lake Chad, where genocidal murders have become commonplace in the region of Darfur. In Niger, just to the west of Lake Chad, the regionwide drought has contributed to the famine conditions that put millions at risk.

There are many complex causes of the famine and genocide, but a little-discussed contributing factor is the disappearance of Lake Chad, formerly the sixth largest lake in the world, in a period of only the last 40 years.

LAKE CHAD, AFRICA

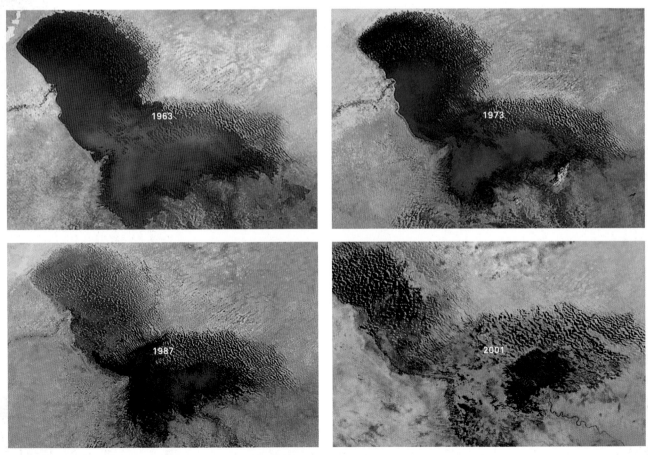

1963

1973

1987

2001

THE EFFECTS OF LAKE CHAD'S DISAPPEARANCE

Just 40 years ago Lake Chad was as large as Lake Erie. But now, due to declining rainfall and ever-intensifying human use, it has shrunk to one-twentieth its original size. Yet today more people rely on Lake Chad than ever, even as sand dunes cover its drying bed. Its fate is sadly emblematic of a part of the world where climate change can be measured not just in temperature increases but in lives lost. The lake's dissipation has led to collapsing fisheries and crops, displacing millions and imperiling many more.

When it was full, Lake Chad was the sixth largest lake in the world, straddling the borders of Chad, Nigeria, Cameroon, and Niger. People depended on its waters for crop irrigation, fishing, livestock, and drinking water. N'guigmi, a city in Niger once surrounded on three sides by Lake Chad, is now more than 60 miles from the water. Fishing boats and water taxis there are permanently stranded. Chad and Malafator, in Nigeria, have suffered similar fates. When Nigerian fishermen followed the receding water into neighboring Cameroon, they sparked military firefights and international legal disputes. When farmers began to till the former lake bottom, battles over property rights erupted.

While Lake Chad withered, periods of particularly intense drought set the stage for the violence that has erupted in neighboring Darfur, a war-torn region of Sudan. To the north and west, Morocco, Tunisia, and Libya each lose some 250,000 acres of productive land a year to the desert. And south in Malawi, 5 million people faced starvation in 2005 when farmers planted on schedule but the rains failed to come. Most Africans still rely literally on the fruits of their labor; when crops fail, things fall apart.

These problems are expected to worsen. Scientists have predicted that by the end of this century, people in many African cities will lose between a quarter and a half of the river flow they rely on to survive. In particularly dry years, some 20 million people could lose the crops that feed them. The famously lush Okavango Delta in Botswana could lose three-quarters of its water, jeopardizing its menagerie of more than 450 bird species and elephants and top predators. Africa's wildlife draws visitors from all over the world and losing it would destroy the leading economic engine of the region, which is tourism.

In the debates that ignite around famine relief, it is sometimes implied that Africans have brought this upon themselves through corruption or mismanagement. But the more we understand about climate change, the more it looks as if we may be the real culprit. The United States emits about a quarter of the world's greenhouse gases, while the entire continent of Africa is culpable for only about 5% of it. Just as we cannot actually see the greenhouse gases, often we do not see their impact from such a long distance. But it is time to take a hard, honest look at our role in this escalating disaster. We helped manufacture the suffering in Africa, and we have a moral obligation to try to fix it.

SUDANESE MOTHER AND HER CHILD IN A FOOD DISPENSARY, KALMA, SOUTH DARFUR, 2005

A second reason for the paradoxical effect of global warming is that while it produces more evaporation from the oceans to fill the warmer atmosphere with increased moisture, it also sucks more moisture out of the soil.

Partly as a consequence, desertification has been increasing in the world decade by decade.

The chart to the right shows the global impact, measured in square miles, per year. The latest figures are significantly worse.

ROAD INTERRUPTED BY A SAND DUNE, NILE VALLEY, EGYPT, 1991

GLOBAL YEARLY DESERTIFICATION

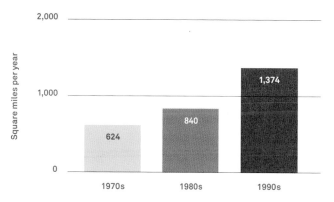

Square miles per year

624	840	1,374
1970s	1980s	1990s

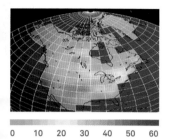

0   10   20   30   40   50   60
Percentage of loss

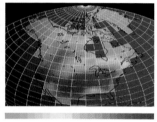

0   10   20   30   40   50   60
Percentage of loss

SOURCE: PRINCETON GFDL R15
CLIMATE MODEL; CO_2 TRANSIENT
EXPERIMENTS

Soil moisture evaporation also increases dramatically with higher temperatures in the United States.

The map to the left shows what is projected to happen to soil moisture in the United States with the doubling of CO_2, which would happen in less than 50 years if we continue business as usual. According to scientists, it will lead, among other things, to a loss in soil moisture of up to 35% in vast growing areas of our country. And of course, drier soils mean drier vegetables, less productive agriculture, and more fires. Moreover, scientists are now telling us that if we do not act quickly to contain global warming pollution, we will soon barrel right through a doubling of CO_2 and move toward a quadrupling, in which case, scientists tell us, most of the United States would lose up to 60% of its soil moisture.

How do we debate something as cataclysmic as this in the traditional framework of our political dialogue?

FARMER IN DROUGHT-RAVAGED
FIELD, WHARTON COUNTY,
TX, 1998

Concrete
and
Countryside

—◆—

I breathed freely—full-chested, invigorating breaths—unlike any I ever took on the streets of Washington, DC

From the time I was born until I went off to college, I had the unusual experience of dividing every year of my life between two radically different landscapes. For eight months each year, because my father was a senator from Tennessee working in Washington, DC, my family's home was a small apartment: No. 809 in the Fairfax Hotel.

My sister, my mother, my father, and I all shared the single bathroom that connected my parents' bedroom to the bedroom I shared with Nancy. The rest of the apartment was made up of a small living room and a dining room with kitchenette attached. The windows looked out on concrete parking lots and other buildings.

The other four months of the year, we lived on a big, sprawling, beautiful farm in Tennessee with animals, sunlight, and grass, all nestled in a sweeping bend of the clear and sparkling Caney Fork River. Alternating between these two settings, year after year, gave me what I now see in retrospect was an unusual opportunity to compare one reality with the other—to compare them not intellectually, but emotionally. Both places were changing and growing during those years, but in different ways, and the city was changing far more rapidly than the rural area surrounding our farm.

Over time, I came to cherish more and more intensely my days on the farm. The scenes may sound cliché now, but in this case, the clichés came to life: soft grass, open sky, rustling trees, cool lakes. I breathed freely—full-chested, invigorating breaths—unlike any I took on the streets of DC.

It's not as if my family didn't have a happy life in that hotel apartment. We did. But the space there was closed and cramped, separated from the natural world. Like millions of families today, I took that separation in stride; I was used to it. I didn't bemoan the fact that our window looked eight stories down to the street.

In fact, when I was very young, I used to climb the fire-escape stairs to the roof with one or another of my friends after school. We tied threads around the necks of plastic soldiers, lowering them spool by spool (from my mother's sewing case) seven-and-a-half stories to bop the top of the doorman's hat and watch him swat at the air. When I was just a little older, my friends and I would throw water balloons from the same perilous location onto the roofs of cars stopped for the red light at 21st Street and Massachusetts Avenue. In other words, I found ways to have fun, though these games, were I my own parent, would absolutely have terrified me.

But the farm was always a different kind of experience. I couldn't wait to get back there; I loved the farm. As a kid, I often walked with my father over every part of the place, learning from him to appreciate the details of the terrain. My dad taught me the moral necessity of caring for the land. He never used those exact words, but that's what his lessons were all about.

He taught me how to recognize the faintest beginnings of a gulley cut by rainwater draining through the churned soil of a plowed field. He showed me how to place rocks or branches in the path of the tiny rivulets, to disperse them and take away their ability to slice through the soil. He made me understand that if the water continued on its way undeterred, it would quickly open deeper gashes, scarring the land and making it infertile by carrying away the rich topsoil.

Such scars were common in the 1920s and early 1930s throughout the South and other parts of America, and my father's instinct to protect and mend the land was painstakingly passed on to me through repeated instructions in measured tones. Had I not learned those lessons early while watching him do the work with his own hands, I might have found them impossibly abstract, with no relevance to me. But to this day, when I walk on what is now my farm with my children and grandchildren, I teach those same lessons and find myself, to my own amazement, using many of the same words I heard when I was 4 and 12 and 20.

From my father, I learned about the duty to care for the land. But it was from my mother that I first learned about the Earth's vulnerability to human harm. When I was 14 years old, my mother read *Silent Spring* by Rachel Carson. And it

made such a big impression on her that she insisted on reading passages aloud after dinner each night for 10 or 12 nights running. My sister and I sat with her at our dining-room table, and we listened. One reason I remember this so well is that Carson's book was the only one my mother ever treated that way. She was always reading books, and when I was very little she read to me a lot. But she stopped after I got older, except for this one book, which was somehow different. I never forgot it.

Silent Spring connected the basic lesson of stewardship I'd learned in my childhood with a new lesson: that human civilization was now capable of seriously harming the environment in ways it simply hadn't been able to in the past. And it was as wrong to ignore this lesson as to walk past a developing gulley on our farm in Tennessee.

So many years spent moving back and forth between Washington and Carthage had its disadvantages, to be sure. But I believe it gave me a perspective on nature—the environment, if you will—that I'm glad I had a chance to learn. Had I grown up entirely on the farm, I think I might well have taken nature much more for granted. But being deprived of it at the end of each summer allowed me to know it by its absence and to better appreciate its incomparable grace. Had I grown up entirely in the big city, I might never have known what I was missing and might never have understood Rachel Carson's warning in a personal, moral context.

When I was elected to Congress in 1976, Tipper and I decided to follow the same pattern with our children that had characterized my own upbringing: school in the big city, every summer and Christmas on the farm.

There are two places on Earth that serve as canaries in the coal mine— regions that are especially sensitive to the effects of global warming. The first, pictured at right, is the Arctic. The second is the Antarctic.

In both of these frozen realms, scientists are seeing faster changes and earlier, more dramatic effects of climate change than anywhere else on Earth.

In photographs, these two ends of the Earth superficially resemble one another. In both places, ice and snow are everywhere you look. But beneath the surface, there is a dramatic difference between them. In contrast to the massive, 10,000-foot-thick Antarctic ice cap, the Arctic ice cap is, on average, less than 10 feet thick. And beneath the ice at each pole lies the reason for the difference: The Antarctic is land surrounded by ocean, while the Arctic is ocean surrounded by land.

The sheer thinness of the Arctic's floating ice—and of the frozen layer of soil in the land area north of the Arctic Circle surrounding the Arctic Sea— makes it highly vulnerable to the sharply rising temperatures.

As a result, the most dramatic impact of global warming in the Arctic is the accelerated melting. Temperatures are shooting upward there faster than at any other place on the planet.

ARCTIC

127

This image shows the largest ice shelf in the Arctic—the Ward Hunt shelf. Three years ago it cracked in half, to the astonishment of scientists. This had never happened before.

RESEARCHERS DISCOVER THE BREAKUP OF THE WARD HUNT ICE SHELF IN NUNAVUT, CANADA, 2002

In Alaska these are called "drunken trees" because they are leaning every which way. And this is caused neither by wind damage nor alcohol consumption.

These trees put their roots deep into frozen tundra decades—even centuries—ago and now as the tundra melts they lose their anchor, causing them to sway in all directions.

The land areas north of the Arctic Circle are frozen for most of the year. Some of the soil that remains permanently frozen is referred to as "permafrost." However, global warming has begun to thaw large areas of permafrost.

That is why the building to the left, located in Siberia, is collapsing. It was built on permafrost and now the permafrost is giving way.

It is the same reason that the house in Alaska pictured below, at left, had to be abandoned by its owner.

INFRASTRUCTURE AT RISK BY 2050 DUE TO PERMAFROST THAW

The Arctic Council just completed a study of infrastructure damage expected from the thawing of frozen tundra around the Northern Hemisphere. Areas marked in pink show where the most severe damage is predicted. Notice the enormous area in Siberia that is affected, approximately 1 million km² of land that has been frozen since the last ice age. According to scientists, this area of tundra contains 70 billion tons of stored carbon, which is becoming unstable as the permafrost melts. The carbon in these Siberian soils is 10 times the amount emitted annually from man-made sources. The leading Russian scientific expert in this field, Sergei Kirpotin from Tomsk State University, issued a grave warning: the permafrost melting is an "ecological landslide...connected to climate warming."

SIBERIA

Stable
Low Risk
Medium Risk
High Risk

SOURCE: ACIA

A TRUCK TRAVELING ALONG
A WINTER ROAD ON THE FROZEN
KOTUY RIVER. TAYMYR,
NORTHERN SIBERIA, RUSSIA, 2004

Trucks that must travel on frozen highways in Alaska for most of the year now sometimes get stuck in the mud as the permafrost thaws.

Ironically, the oil companies that are trying to convince the U.S. Congress to let them drill for oil in protected areas on the northern slope of Alaska would also have to depend on frozen highways. But now, the massive thawing of the permafrost further complicates their already controversial proposal.

The graph below shows the number of days each year that the tundra in Alaska is frozen solidly enough to drive on.

Currently the number has fallen to fewer than 80 days per year. Spring is coming earlier; fall is arriving later. And all the while, the temperature keeps going up—more rapidly in the Arctic than anywhere else in the world.

ALASKA WINTER TUNDRA TRAVEL DAYS: 1970–2002

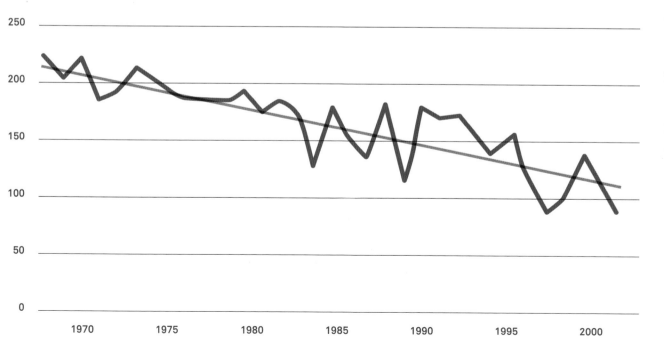

The nation's pipeline is in trouble due to melting permafrost.

A PRIMER ON ALTERNATIVE FUELS

When I was in Congress we used to wrangle about the value of making ethanol (that's grain alcohol) from corn. Despite the moonshine jokes, I supported ethanol. Even though some of its environmental consequences made me uncomfortable, I thought it was important for us to work on alternatives to fossil fuels—to begin to break our dependence on foreign oil. And I'm happy to report that since then, even newer innovations have begun to make a difference.

A Canadian company has figured out a way to make a new kind of ethanol out of plant fiber—meaning that it's cheaper and cleaner than regular ethanol. It's called cellulosic ethanol, which simply means that instead of being made from the sugar in corn it's made from cellulose, a tough plant fiber. Cotton, for example, is almost pure cellulose, and most agricultural waste is loaded with it. So now we're able to use cornstalks instead of corn. And other low-maintenance, high-yield plants like switchgrass and poplar can also be cheaply converted to alcohol. By one estimate, this new technology means that crop waste could create 25% of the energy needed for transportation. And while ethanol from corn creates 29% less greenhouse gas than gasoline, ethanol from cellulose could cut gases by 85%.

Biodiesel is another alternative fuel you may have heard of—it can be brewed from leftover frying oil. If you have ever wondered how many french fries we would need to eat to really break OPEC, Changing World Technologies is making biodiesel without the Crisco. The company has perfected a bioreactor that can take anything organic and turn it into oil. The first production facility adjoins a Butterball factory in Carthage, Missouri. Take 270 tons of turkey guts, add 20 tons of pig fat, process them, and you end up with 500 barrels of high-grade biodiesel. Sewage, used tires, and plastic bottles can now be transformed into fuel.

Hydrogen may be the ultimate clean fuel of the future. But most experts agree that a hydrogen economy is at least a few decades away. We also know that what works in one part of the country may not in another. For instance, solar-produced hydrogen in Arizona is sensible—and feasible—because that state averages 300 days of sunshine per year. But cracking the hydrogen out of coal or natural gas produces a stream of almost pure carbon dioxide, which—unless locked away— could make the greenhouse effect worse.

Part of our future work is to make the right decisions in the right places—to know what is realistic for each state, each ecological and industrial system. We want to boost local economies wisely and responsibly without worsening the trends that have brought us to a point of deep crisis.

From Pole to Shining Pole

◆·◆

One can read field studies, talk to scientists, and scrutinize charts, but there's nothing like seeing things for yourself.

The story I have tried to tell about global warming is a story that involves a double journey: one metaphorical and one real. The slideshow that I give regularly to audiences around the world is the distillation of my own intellectual journey toward understanding the nature of this crisis and our difficulty in facing up to it. Every part of the slideshow replicates an "Aha!" moment that I have had in the process of my own education. My goal in presenting the slideshow is to recreate that moment of discovery for others.

But my own journey toward understanding the crisis has also involved a literal journey to many hard-to-reach places where global warming is in clear evidence, remote pockets of the planet where many of the world's best scientists are frequently hard at work, often under extremely difficult conditions.

One can read field studies, talk to scientists, and scrutinize charts, but there's nothing like seeing things for yourself. I have been drawn to these excursions not only because of my hunger to learn as much as I can about the climate crisis, but also in part, I think, because they take me outdoors again and give me a chance to see many truly fascinating places.

Exploring hot spots where global warming has left its imprint, I've traveled from the Greenland Ice Dome to the swamps of the Everglades; from the Aral Sea to the Dead Sea; from the North Slope of Alaska to the South Island of New Zealand; from the Serengeti Plains to the Kyzyl Kum Desert; from the Nile to the Congo; from the Skeleton Coast of Namibia to the Galapagos Islands; from Mauna Loa to the Mekong Delta; from the Badlands to the Cape of Good Hope; from Oak Ridge

National Laboratory in Tennessee to the Chernobyl Sarcophagus in Ukraine; from the Amazon Jungle to Glacier National Park; from the highest lake, Titicaca, to the lowest desert, Death Valley. But of all the places I've visited, the North Pole and the South Pole truly stand out.

When I went to the South Pole, several things surprised me. First, the ice-and-snow pack there is more than 10,000 feet thick—so I experienced the same altitude sickness that most first-time visitors feel: a slight headache and nausea that soon pass as one acclimates to the height. It didn't occur to me beforehand that the average altitude in Antarctica is far higher than the average altitude on any other continent of the planet; the ice and snow have piled layer upon layer for so many hundreds of thousands of years that they've pushed up the top of the ice

Gore aboard Marine II, on a trip to Glacier National Park, 1997

cap high into the sky. And geologists tell me that the weight of the ice and snow has pushed down the bedrock underneath it to below sea level.

Also, since the precipitation there is so limited, each layer of snow is quite thin, presenting a challenge for scientists seeking data from the layers. The oldest layers near the bottom have been compressed by the weight of the stack, further complicating the task of measuring CO_2 content in the tiny, trapped air bubbles.

Third, though I knew it would be cold in Antarctica, I really had no idea how cold. The forecast said "58° below zero," but nothing in my prior experience equipped me to understand what that would mean. And that's when I learned a particularly interesting lesson from one veteran of several seasons at the South Pole Station:

"There is no such thing as bad weather," he said, "just bad clothes."

Understandably, the scientists who live and work in Antarctica pay a lot of attention to clothes (though very little attention to how the clothes look). The hoods of their specially designed parkas extend far out in front of their faces, because the air is so cold that it must be warmed up—at least a little—before it's breathed in. And most people cannot expose their heads for very long without risking serious frostbite to their ears. As a result, people walk around peering at the outside world through a foot-long tunnel of thick rabbit fur.

At the exact location of the South Pole, there is actually a barber pole stuck in the ice. This serves two purposes: it allows visitors to "run all the way around the world," and it is the place where visitors

take pictures. The recommended technique for photos, by the way, is to throw off the hood for a few seconds, smile bravely while the picture is snapped, then quickly pull the hood back over the head.

Yet another surprise for me was when scientists showed me that near the South Pole, the presence of air pollution in the ice cores visibly declined not long after passage of the U.S. Clean Air Act in 1970; looking back through the annual layers of ice, you can actually see the before and after with your own eyes. One thing both Antarctica and the Arctic have in common is their remoteness from civilization. Yet both of these formerly pristine locations are now marked by industrial pollution. The air above the North Pole contains pollution at levels still rising because of the prevailing wind patterns

in the Northern Hemisphere and the higher concentration of industry in that hemisphere.

I first visited the Arctic ice cap only two and a half years after visiting Antarctica, and this drove home the startling contrasts. I flew to Deadhorse, Alaska, on the shores of the Arctic Ocean and then traveled by helicopter to a rendezvous with a submarine for the journey due north under the ice cap.

On my second trip, I flew to Greenland, where I switched first to a C130 specially equipped with skis, then to a smaller ski-equipped plane. After a three-and-a-half-hour flight due north in the small plane, we landed on an ice floe in the Arctic Ocean and switched to skimobiles.

We stopped for a few hours of sleep in tents set up on the ice, then remounted the skimobiles for a two-mile run to the northern edge of the floe. There, Navy corpsmen used brooms to mark a giant X on the spot where a submerged submarine was expected to surface—and we backed away to what we thought was a safe distance.

I watched in awe as the giant submarine crashed upward through the ice. But as it rose, a crack shot outward from the break like a lightning bolt, straight toward me. Instantly, I dove to one side of the fis-

sure. As I got to my feet, I noticed the nearby corpsmen smiling; their reaction to the splitting ice had been considerably calmer than mine.

After resubmerging, we traveled seven hours due north, all the way to the Pole. When we arrived, the navigational display lined up a row of zeros that made me feel like we had won the jackpot on a high-tech slot machine. We circled around and surfaced exactly at the Pole.

I remember climbing out of the conning tower and standing on the ice. What struck me most was the beautiful, almost magical quality of the tiny ice crystals in the air around me, reflecting bright sunlight like airborne jewels.

The reason for my two trips under the Arctic ice cap was to learn more about global warming and to convince the U.S. Navy to release top-secret data to the environmental scientists specializing in the study of global warming's effects on Arctic ice.

The reason the Navy has collected data on ice thickness that no one else has is that they've regularly patrolled the Arctic Ocean—under the ice—for almost 50 years in a fleet of specially designed submarines capable of surfacing through the ice. During the Cold War, our strategic military forces were poised to retaliate on a few minutes' notice in the event of an attack by the former Soviet Union.

So the submarines in the Arctic had to be capable of surfacing on short notice in the event of a nuclear exchange.

But that posed a challenge, because the ice cap is much thicker in some places than in others. Even with their special design, these submarines can surface safely only in spots where the ice is three feet thick or less. So the Navy has long used a special upward-looking radar to measure the thickness of ice above the subs as they make their long journey under the ice cap. And over the past half-century, the Arctic submarines have maintained a careful record of ice thickness for each of the "transects," or trips under the ice.

It was that five-decade record of ice thickness, marked "top secret," that I was after. I wanted the Navy and the CIA, which also had authority over the data, to release this one-of-a-kind record to the scientists who desperately needed it to answer the crucial questions about global warming: Is the North Polar ice cap melting? And if so, how fast?

At first, the Navy strenuously resisted any release of the data, fearing it might be of assistance to enemies of the United States who could use it to figure out the submarine patrol routes. As a member of

the Senate Armed Services Committee, I completely understood, and I worked with the Navy to reconcile their legitimate concern with the environmental imperative that was also of critical importance. Bruce DeMars, the four-star admiral who then headed the Navy's Nuclear Propulsion Program and was responsible for all the submarines, actually went with me to the North Pole, listened to me talk about global warming the entire trip, and, despite his initial skepticism, became an invaluable ally. He and President Reagan's CIA director, Bob Gates, were the ones who came up with an innovative solution allowing release of the data under careful safeguards.

It was a good thing they did, because the information proved to be even more significant and alarming than the scientists had expected: It showed a dramatic pattern of rapid melting. In recent decades, the submarine data has been combined with satellite imagery to give an even more comprehensive picture that shows the North Polar ice cap began a fairly rapid retreat in the mid-1970s.

While the impact of global warming is most pronounced in the Arctic and the Antarctic, I found dramatic evidence of

significant effects at the Equator as well.

During two trips to the Amazon rainforests, I found scientists there becoming even more deeply concerned about dramatic shifts in rainfall patterns. In 2005 the Amazon suffered the longest and worst drought in recorded history —with devastating effects.

In Kenya, also on the Equator, I heard growing concerns about the increased threat from mosquitoes and the diseases they can transmit in higher altitudes that were formerly too cold for them to inhabit.

In all of my journeys, I have searched for a better understanding of the climate crisis—and in all of them I have found not only evidence of the danger we face globally, but an expectation everywhere that the United States will be the nation to lead the world to a safer, brighter future. And as a result, since every journey took me back home, I have returned each time with a deeper conviction that the solution to this crisis that I have traveled so far to understand must begin right here at home.

Gore on the Equator in Africa, 1989

On two occasions, I traveled underneath the Arctic ice cap in a nuclear submarine, which then surfaced through the ice. The second time it surfaced precisely at the North Pole. The series of photographs below shows one of the U.S. Navy's specially designed Arctic subs that have been patrolling under the ice cap for almost 50 years, continually since the first mission of the USS Nautilus in 1958. Notice the wings on the submarine's conning tower; on most subs, they remain in a horizontal position to help steer the boat as it glides through the water. On these models, however, they can be rotated to the vertical position (as seen below) in order to function as blades to assist in breaking through the ice as the sub is surfacing. Since they can surface only in areas where the ice is three feet thick or less, the Navy has kept a meticulous record of ice thicknesses measured by upward-looking radar.

**VIDEO GRABS OF USS PARGO
SURFACING THROUGH ICE**

This data was considered classified by the Navy for many years. When I persuaded them to release it, the data told an alarming story.

Since the 1970s, the extent and thickness of the Arctic ice cap has diminished precipitously. There are now studies showing that if we continue with business as usual, the Arctic ice cap will completely disappear each year during the summertime. At present, it plays a crucial role in cooling the Earth. Preventing its disappearance must be one of our highest priorities.

SEA-ICE EXTENT: NORTHERN HEMISPHERE

SOURCE: HADLEY CARTER

The reason this Arctic ice cap has been melting so quickly is first because it is much thinner than the Antarctic ice cap, since it floats on top of the Arctic Ocean. Second, as soon as a portion of the ice melts, there is a dramatic difference in the amount of heat absorbed from the sun. As seen in the illustration at right, the ice reflects most of the incoming solar radiation, like a giant mirror, whereas the open sea water absorbs most of that heat. As the water warms up, it puts even more melting pressure on the edge of the ice adjacent to it. It is an example of what scientists call "positive feedback," and it is happening right now in the Arctic.

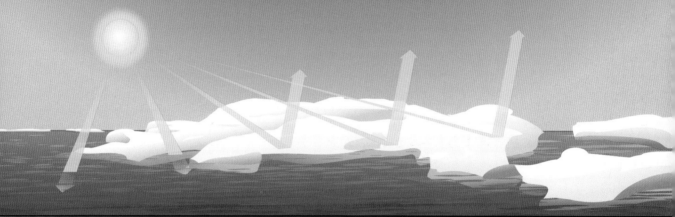

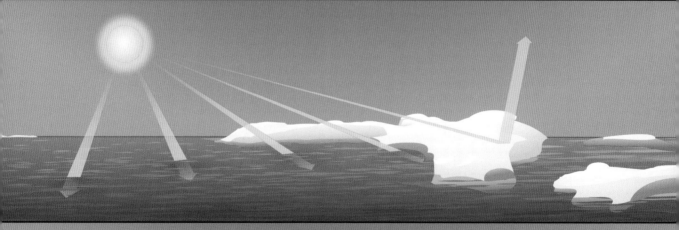

The melting of the ice represents bad news for creatures like polar bears. A new scientific study shows that, for the first time, polar bears have been drowning in significant numbers. Such deaths have been rare in the past. But now, these bears find they have to swim much longer distances from floe to floe. In some places, the edge of the ice is 30 to 40 miles from the shore.

What does it mean to us to look at a vast expanse of open water, at the top of our world, that used to be—but is no longer—covered by ice? We ought to care about this a lot, because it has serious planetary effects.

MOTHER AND POLAR BEAR CUB ON PACK ICE, SPITZBERGEN, NORWAY, 2002

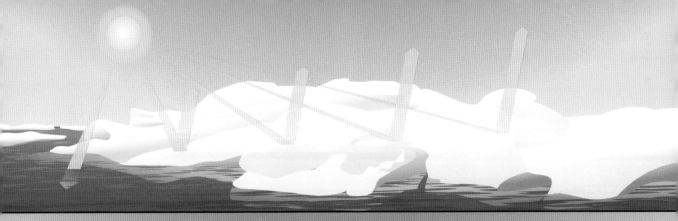
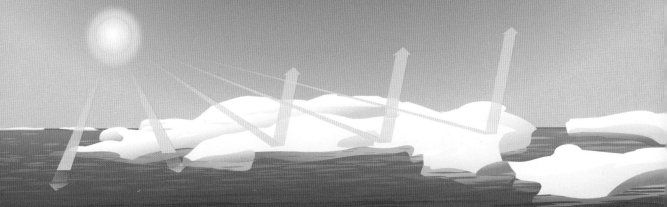

The melting of the ice represents bad news for creatures like polar bears. A new scientific study shows that, for the first time, polar bears have been drowning in significant numbers. Such deaths have been rare in the past. But now, these bears find they have to swim much longer distances from floe to floe. In some places, the edge of the ice is 30 to 40 miles from the shore.

What does it mean to us to look at a vast expanse of open water, at the top of our world, that used to be—but is no longer—covered by ice? We ought to care about this a lot, because it has serious planetary effects.

**MOTHER AND POLAR BEAR CUB
ON PACK ICE, SPITZBERGEN,
NORWAY, 2002**

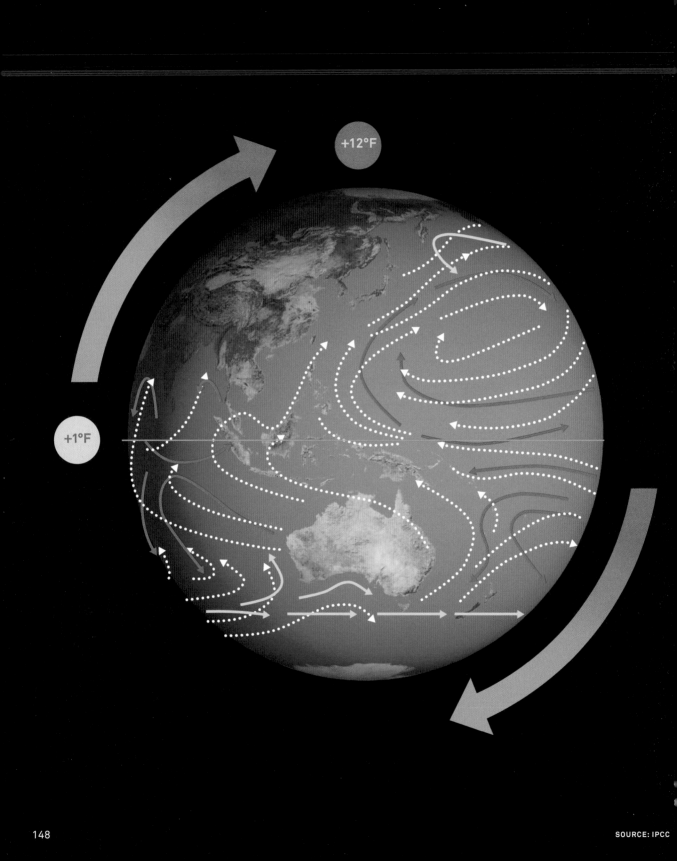

+12°F

+1°F

SOURCE: IPCC

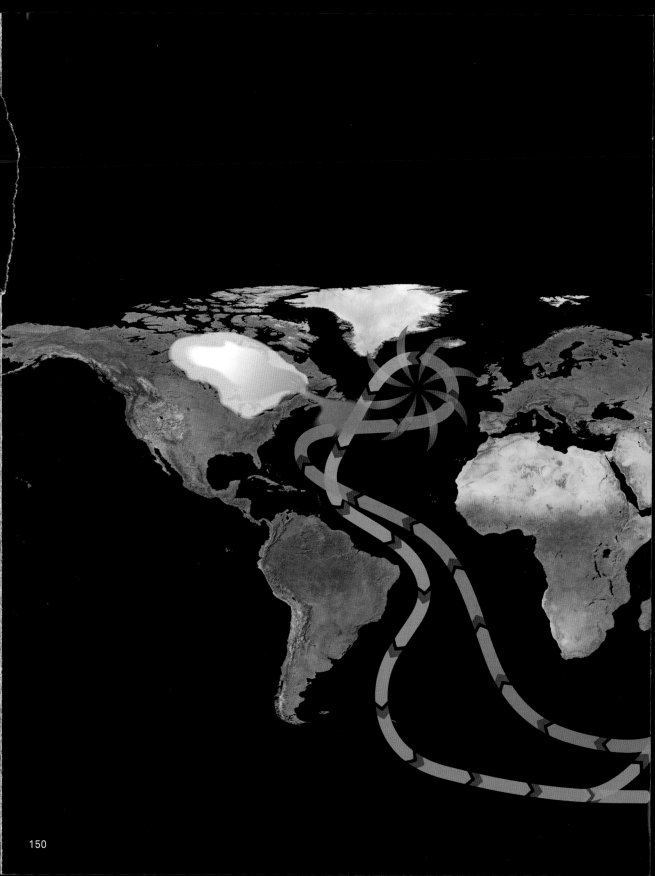

Melting the Arctic could profoundly change the planet's entire climate pattern. Scientists call the global climate a "nonlinear system," which is just a fancy way the scientists have of saying that the changes are not all gradual. Some of them can, and have in the past, come suddenly, in big jumps.

These scientists say that the world's climate is best understood as a kind of engine for redistributing heat from the Equator and the tropics to the poles. Much more solar energy is absorbed by the Earth between the Tropic of Cancer and the Tropic of Capricorn because the Sun is directly overhead every day all year long.

By contrast, the Sun's rays strike only glancing blows at the North Pole and the South Pole. Each receives the sunlight for only half the year, during which the other is completely in darkness.

The redistribution of heat from the Equator to the poles drives the wind and ocean currents—like the Gulf Stream and the jet stream. These currents have followed much the same pattern since the end of the last ice age 10,000 years ago, since before the first human cities were built. Disrupting them would have incalculable consequences for all of civilization. And yet, the climate crisis is gaining the potential to do just that.

The average temperature worldwide is about 58°F.

An increase of five degrees actually means an increase of only one or two degrees at the Equator, but more than 12° at the North Pole, and a large increase on the periphery of Antarctica as well.

And so all those wind and ocean current patterns that formed during the last ice age, which have been relatively stable ever since, are now up in the air.

Our civilization has never experienced any environmental shift remotely similar to this. Today's climate pattern has existed throughout the entire history of human civilization.

Every place—every city, every farm— is located or has been developed on the basis of the same climate patterns we have always known.

According to scientists, one surprisingly fragile component of the global climate system is in the North Atlantic, where the Gulf Stream encounters the cold winds coming off the Arctic and across Greenland. As the two collide, heat evaporates out of the Gulf Stream and is carried as steam by the prevailing winds and the Earth's rotation eastward to Western Europe.

The currents of the ocean are all linked like a big Möbius strip in a loop called "The Global Ocean Conveyer Belt." The red parts of the loop below represent the warm surfaces, the best known of which is the Gulf Stream, which flows along the east coast of America. The blue portions of the loop represent the deep cold-water currents flowing in the opposite direction.

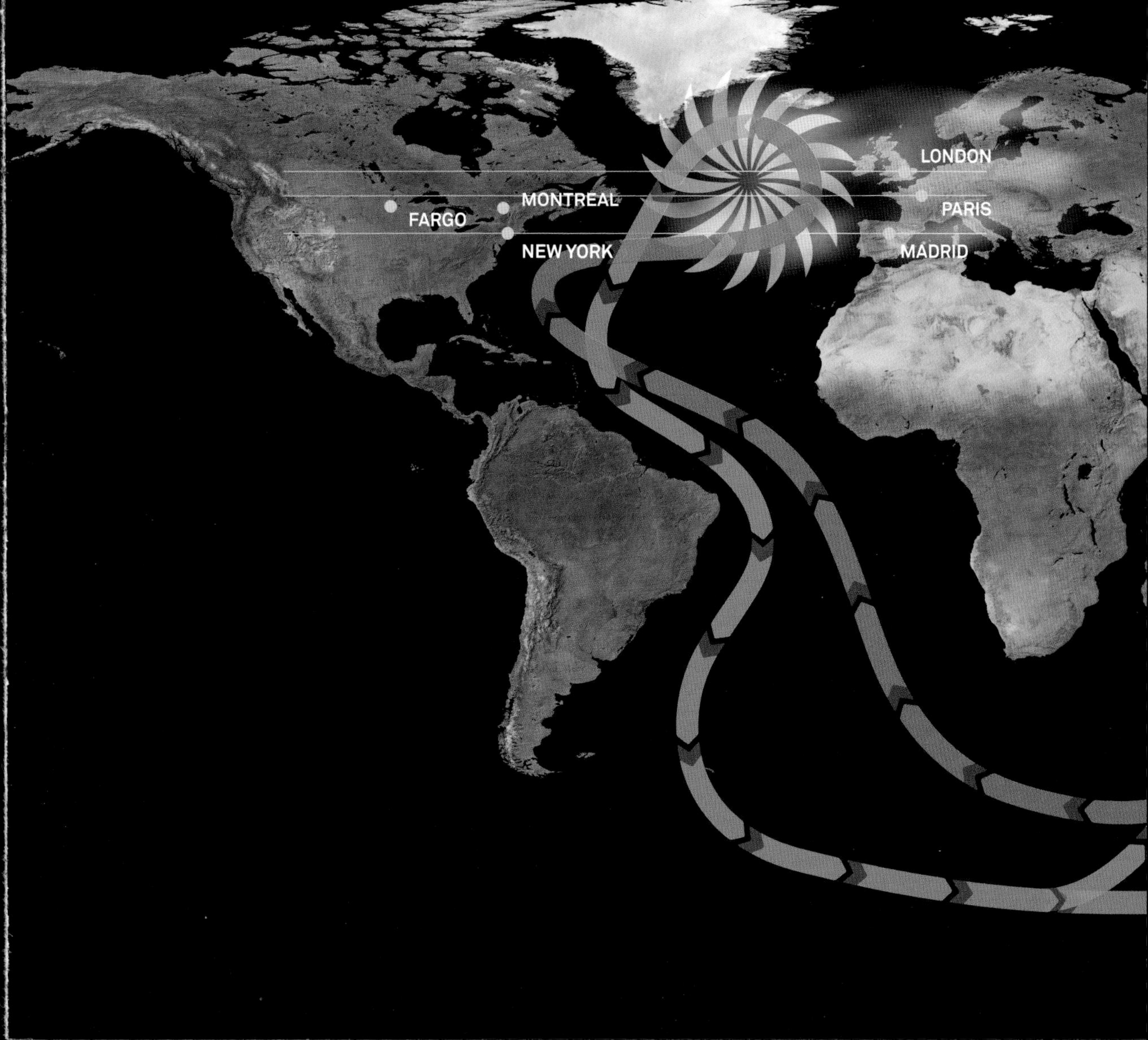

LONDON

PARIS

FARGO

MONTREAL

MADRID

NEW YORK

This massive sinking phenomenon is described by scientists as a giant pump. More specifically, they call it a "thermohaline pump" because it is driven both by temperature (thermo) and by salinity (haline). This pump plays a crucial role in powering the continuous flow of the world's ocean current system.

Around 10,000 years ago, something happened that the scientists are worried could happen again. When the last glacial ice sheet in North America melted, it formed a giant pool of fresh water. The Great Lakes are the remnant of that huge freshwater lake, which was held in place on its eastern boundary by an enormous ice dam.

Then one day the ice dam broke and the fresh water rushed out into the North Atlantic. When that unprecedented amount of fresh water tore open the St. Lawrence River and flooded into the North Atlantic, the pump began to turn itself off. The Gulf Stream virtually stopped. So Western Europe no longer received all of that heat from the evaporating Gulf Stream.

Consequently, Europe went back into an ice age for another 900 to 1,000 years. And the transition happened fairly quickly.

Some scientists are now seriously worried about the possibility of this phenomenon recurring. At Woods Hole Research Center, Dr. Ruth Curry is especially concerned about the rapid melting of ice in Greenland, which is adjacent to the area in which the pump operates.

Recently, she observed: "The possibility of such extreme events precludes ruling out that disruption of the North Atlantic Ocean Conveyor in the 21st century could occur as a result of greenhouse warming."

Incidentally, the heat drawn from the Gulf Stream and carried to Europe makes cities like Paris and London much warmer than Montreal or Fargo, North Dakota, even though they are close in latitude. Madrid is much warmer than New York City, though it is on the same latitude.

The water left behind in the North Atlantic when the warm water evaporates is not only colder but also saltier, because all the salt stays right where it is, but in a higher concentration. So the water is much heavier and therefore sinks at the astonishing rate of 5 billion gallons a second. As it drops straight down toward the bottom of the ocean, it forms the beginning of the cold-water current flowing southward.

The age-old rhythm of the Earth's seasons—summer, fall, winter, and spring—is also changing as some parts of the world heat up more rapidly than others.

SHIFTS IN SEASONS: 1980

Bird Arrival   Bird Hatching   Caterpillar Hatching

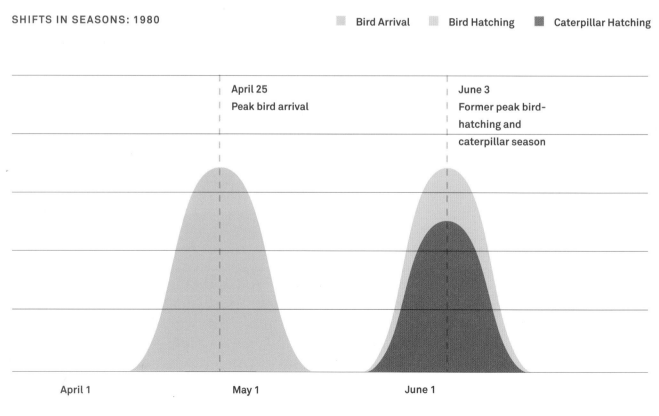

April 25
Peak bird arrival

June 3
Former peak bird-
hatching and
caterpillar season

April 1   May 1   June 1

A study from the Netherlands, depicted below, shows that 25 years ago, the peak arrival date for the migratory birds was April 25. Their chicks hatched almost six weeks later, peaking on June 3, just in time for the height of the caterpillar season. Now, two decades of warming later, the birds still arrive in late April, but the caterpillars are peaking two weeks earlier, leaving the mother birds without their traditional source of food for the chicks. The peak hatching date has moved slightly forward, but cannot move by much. As a result, the chicks are in trouble.

Global warming is disrupting millions of delicately balanced ecological relationships among species in just this way.

BLACK TERN FEEDING YOUNG, DE WIEDEN, NORTHWEST OVERIJSSEL, NETHERLANDS

SHIFTS IN SEASONS: 2000

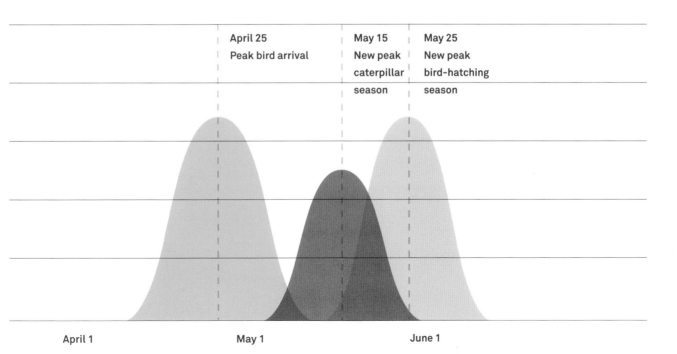

April 25
Peak bird arrival

May 15
New peak caterpillar season

May 25
New peak bird-hatching season

April 1   May 1   June 1

SOURCE: *SCIENTIFIC AMERICAN*

Here is another example of how global warming disrupts the balance of nature as we have known it.

The blue line below plots the sharp decline in the number of days per year with frost on the ground in southern Switzerland. The orange area shows the simultaneous sharp increase in the number of invasive alien species that have rushed in to fill newly created ecological niches.

The same thing is happening here in the United States, too. In the American West, for example, the destructive spread of pine beetles used to be slowed by colder winters that reduced their numbers seasonally. But now, with fewer days of frost, the pine beetles are thriving and the pine trees are being devastated.

SHIFTS IN SEASONS

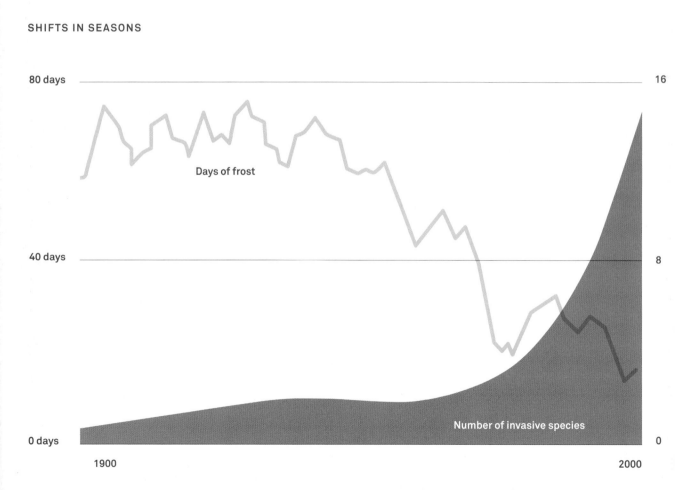

Days of frost

Number of invasive species

80 days — 16

40 days — 8

0 days — 0

1900   2000

SOURCE: *NATIONAL GEOGRAPHIC MAGAZINE*

PINE BEETLE DAMAGE,
PLAINS, MT, 1989

This picture shows a portion of 14 million acres of spruce trees in Alaska and British Columbia that have been killed by bark beetles, whose rapid spread was once slowed by colder and longer winters.

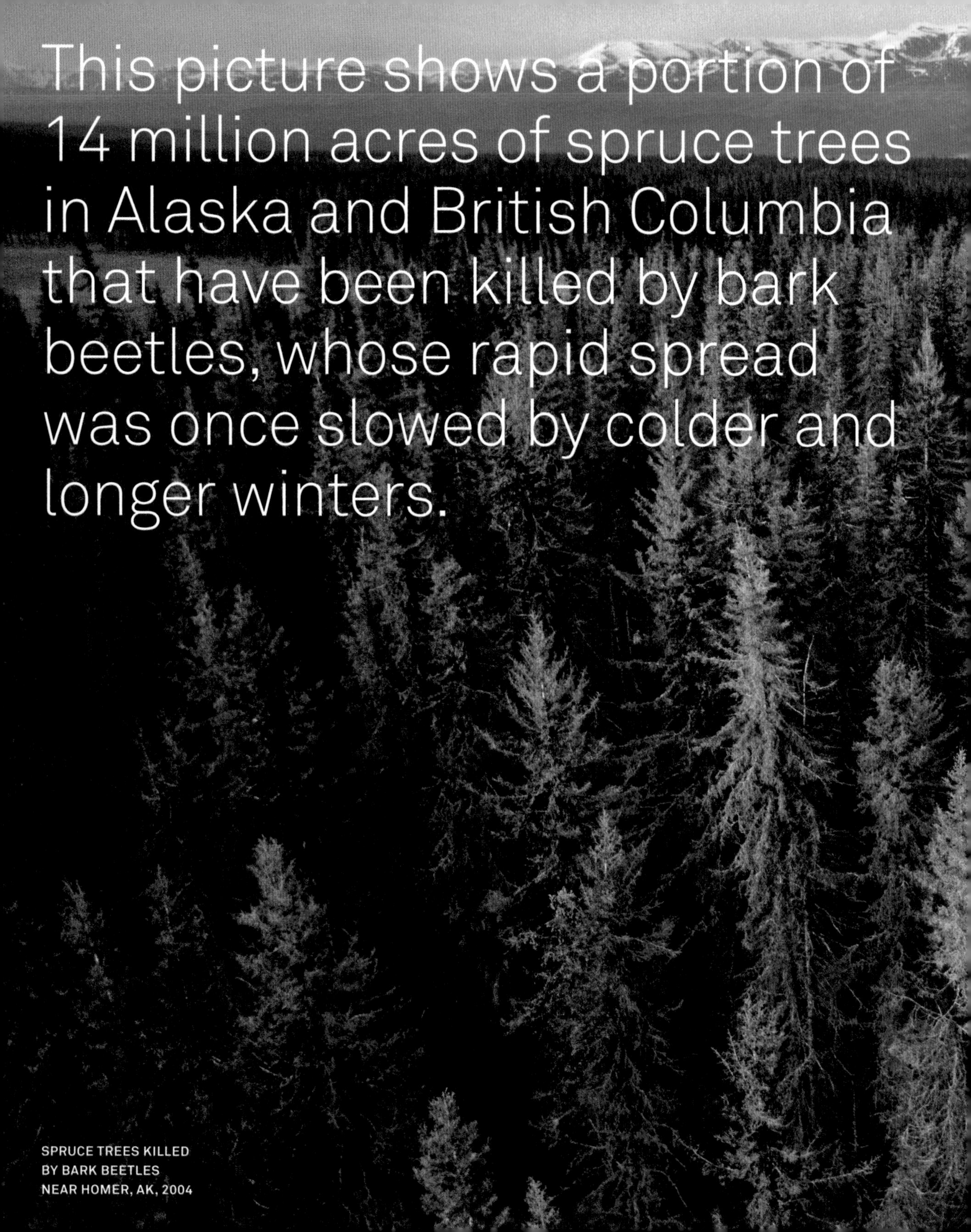

SPRUCE TREES KILLED
BY BARK BEETLES
NEAR HOMER, AK, 2004

Across the Wilderness

◆•◆

**Losing something is one thing;
forgetting what you've lost
is something else again.**

When I returned from Vietnam in 1971, my wife, Tipper, and I bought a tent, a Coleman stove, a lantern, and two backpacks. Then we threw them in the trunk of our Chevrolet Impala and drove across the country, from Nashville to California and back again, camping out all along the way.

We started out by heading north, first to Michigan, then through the Upper Peninsula to Wisconsin and through Minnesota to South Dakota, hopscotching from one national park, national forest, or national monument to the next. I especially remember the Badlands, where we pitched our tent in the primitive campground and took long hikes through the stark, bleak, and barren landscape that conjured images of what we imagined the surface of the Moon to be.

We headed next to Mt. Rushmore, to Devil's Tower, and then on to the majestic sights of Yellowstone National Park and the Grand Tetons. We trekked to the Great Salt Lake and across the Donner Pass, through the High Sierra to Muir Woods and the giant redwoods, ending up on the Pacific shore in Marin County, north of San Francisco.

On the way back home to Tennessee, we saw Yosemite and the Grand Canyon, then Mesa Verde and Santa Fe. It was a wonderful trip, an adventure that dramatized both the breadth of the country and our own small place within it. Tipper and I still marvel at everything we saw and the perspective it offered. At a moment when the country was still in the throes of the conflict over Vietnam, it was refreshing to see the best of America. It was also a private relief to be together again following my long absence so soon after our marriage in 1970.

The following year, we were back in the Impala, camping again, this time exploring Colorado's Rocky Mountains.

Al and Tipper in the High Sierra,
Soda Springs, CA, 1971

and listening to each other. There is a special peace, for me at least, in the simplicity of times like those.

I was fortunate to marry a woman who, among her many endearing qualities, appreciates nature as much as I do. Tipper has often described the same sense of renewal that I feel when we're out on a lake or climbing on rocks. When I'm back in nature after months of walking around on concrete and living in boxes, I feel a palpable internal shift. It doesn't happen right away; I have to settle into it. Sometimes it takes a while to shake off the urban frenzy. But inevitably, that serenity—that stillness—takes hold, and when it comes at last, it's like taking a deep breath and saying, "Oh yeah. I forgot about this."

Losing something is one thing; forgetting what you've lost is something else again. Maybe I shouldn't generalize from my personal experience, but I do believe that our civilization has come perilously close to forgetting what we've lost and then forgetting that we've lost it. This is caused in part by never having the chance to commune with nature. That may sound like so much hippie pap, but I defy anyone to take in this country's unspoiled treasures and not feel calmed, humbled, and rejuvenated by them.

I believe that when God created us (and I do believe evolution was part of the process God used), He shaped us, breathed life and a soul into us, and then set us free within nature, not separate from it, giving

Tipper and I have always shared the same impulse to get outside, to get away, to spend unscheduled time in untamed places. Our vacations have usually involved houseboats, backpacks, or tents.

When our children were old enough, we went back to the Grand Canyon and took all four of them 225 miles down the Colorado River, spending 13 days and nights rafting and hiking by day and sleeping on the riverbank at night. It was exhilarating to spend that kind of uncluttered time together. We climbed into remote parts of the canyon during the midday heat and built campfires at night, swapping stories about our exploits going through the rapids.

Those were among the most fun days we've ever had as a family—all six of us, in sleeping bags, cooking what we ate us intimate connections to all aspects of it. The relationship we have to the natural world is not a relationship between "us" and "it." It is us, and we are of it.

Our capacity for consciousness and abstract thought in no way separates us from nature. Our capacity for analysis sometimes leads us to an arrogant illusion: that we're so special and unique that nature isn't connected to us. But the fact is, we're inextricably tied.

I know there are many people who dismiss the environment as irrelevant to their everyday existence, and I think I know one reason why. When I was in Washington as a kid, I too became addicted to the pace and hum of that kind of life. I sometimes missed Washington's pulse when I returned to Carthage every summer.

Partly as a result, I have a healthy respect for the mesmerizing power of an overscheduled, overpopulated, hyperstimulated existence. It's designed to monopolize our attention, to sell us things, to speed us from one place to the next, to focus us on matters that appear to be vital, even when they're not. It's such an encompassing artificial environment that it can seem to be all there is.

Nature, by contrast, is slow-moving, undemanding, maybe underwhelming for many people. But if you never put

yourself in the midst of nature—to understand that its essence is our essence—then you're inclined to treat it as trivial. You become willing to abuse and destroy it through carelessness, not recognizing that to do so is wrong. Nature becomes the wallpaper of experience, with no deeper meaning in and of itself.

We've come to accept the dominant attitude that if nature can yield something of value to the lucrative engines of commerce, then we should, by all means, grab it and rip it out, never thinking twice about the wounds left behind.

According to this way of thinking, if exploitation results in injury to the environment, so be it; nature will always heal itself, and no one should care.

But what we do to nature we do to ourselves. The magnitude of environmental destruction is now on a scale few ever foresaw; the wounds no longer simply heal themselves. We have to act affirmatively to stop the harm.

TOP: *Al and Tipper hiking, Soda Springs, CA, 1982;*
LEFT: *Gore family on the Colorado River, going down the Grand Canyon, 1994*

GIANT GLASS FROG

GREATER MOUSE LEMUR

WHITE-FRONTED GOOSE

BOWHEAD WHALE

GREY-HEADED ALBATROSS

EMPEROR PENGUIN

GOLDEN TOAD

MACARONI PENGUIN

COQUI (TREE FROG)

FLIGHTLESS CORMORANT

ANTARCTIC FUR SEAL

WATTLED CRANES

YELLOW-EYED PENGUIN

POLAR BEAR

RED-BREASTED GOOSE

LEOPARD SEAL

Many species around the world are now threatened by climate change, and some are becoming extinct—in part because of the climate crisis and in part because of human encroachment into the places where they once thrived.

In fact, we are facing what biologists are beginning to describe as a mass extinction crisis, with a rate of extinction now 1,000 times higher than the normal background rate.

Many of the factors contributing to this wave of extinction are also contributing to the climate crisis. The two are connected. For example, the destruction of the Amazon rain forest drives many species to extinction and simultaneously adds more CO_2 to the atmosphere.

SPECIES LOSS

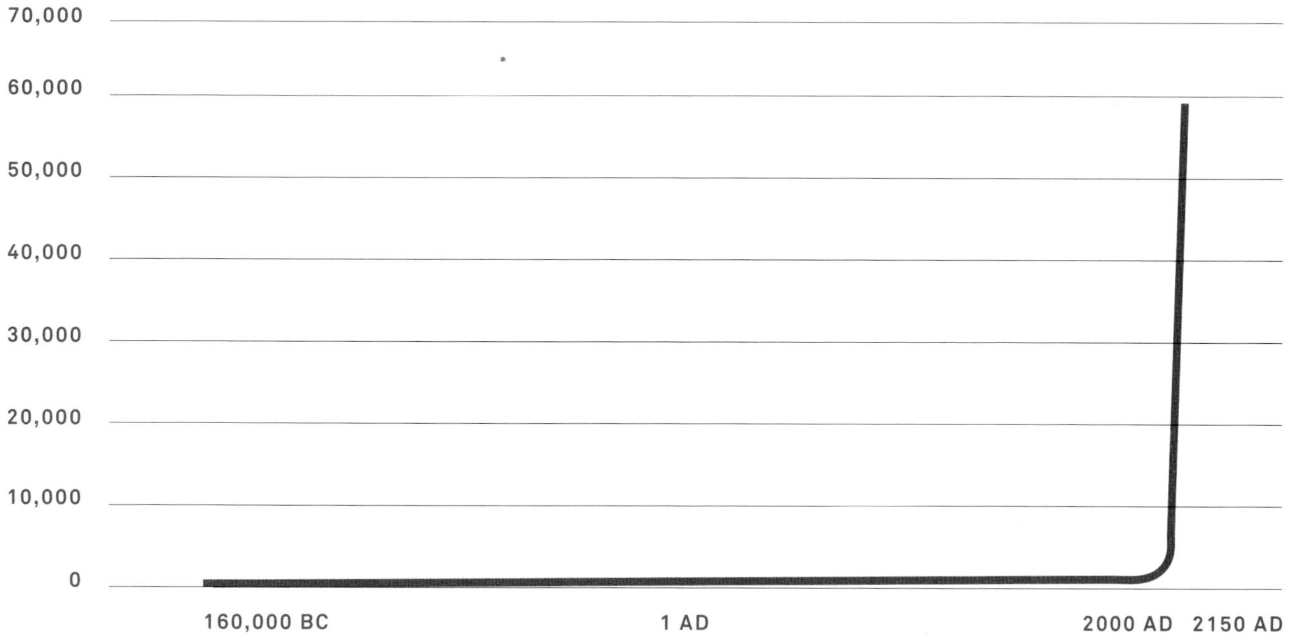

70,000	
60,000	
50,000	
40,000	
30,000	
20,000	
10,000	
0	

160,000 BC   1 AD   2000 AD   2150 AD

SOURCE: UNITED NATIONS

Coral reefs, which are as important to ocean species as rainforests are to land species, are being killed in large numbers by global warming.

In 2005, to date the hottest year on record, there was a massive loss of coral reefs, including some that were healthy and thriving when Columbus first arrived in the Caribbean. In 1998, the second hottest year on record, the world lost an estimated 16% of all its coral reefs.

Many factors contribute to the death of coral reefs—pollution from nearby shores, destructive dynamite fishing in less developed regions, and more acidic ocean waters. However, the most deadly cause of the recent, rapid, and unprecedented deterioration of coral reefs is believed by scientists to be higher ocean temperatures due to global warming.

Coral bleaching—the process that turns healthy, multicolored coral reefs into white or gray skeletons—occurs when tiny organisms living in the transparent membrane covering the skeleton are stressed by heat and other factors and evacuate. When they escape, the thin, clear skin—no longer filled with brightly colored zooxanthellae, or "zooks" as the scientists call them—reveals the colorless calcium carbonate skeleton beneath. The bleached appearance is usually a prelude to the death of the coral.

LETTUCE CORAL,
PHOENIX ISLANDS, KIRIBATI,
POLYNESIA, 2004

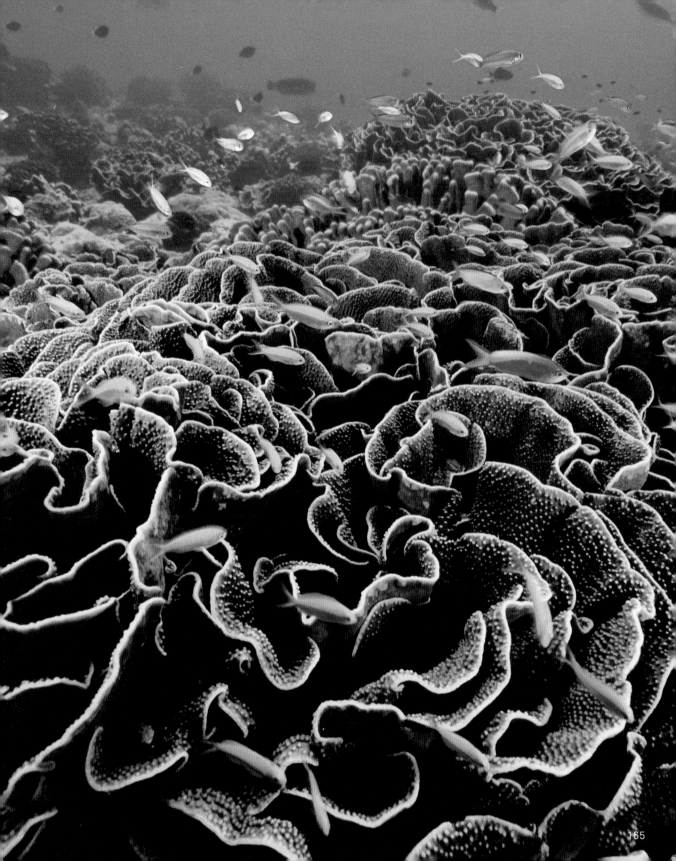

The link between global warming and the large-scale bleaching of corals, considered controversial only 10 to 15 years ago, is now universally accepted.

BLEACHED CORAL, RONGELAP
REEF, MARSHALL ISLANDS, 2004

Corals—along with many other ocean life forms—are threatened by the unprecedented growth of carbon dioxide emissions worldwide, not only because these gases build up in the atmosphere of our planet and increase ocean temperatures, but also because up to one-third of all those emissions end up sinking into the ocean and increasing the acidity of the water. Scientists have recently measured this harmful development.

We are used to thinking about the harmful effects of all the extra CO_2 we've been dumping into the atmosphere. But we now have to worry about the chemical transformation of the oceans as well.

Carbonic acid resulting from all of the extra CO_2 changes the pH in ocean water and alters the ratio of carbonate and bicarbonate ions. This, in turn, affects the saturation levels of calcium carbonate in the oceans—and that is important because many other small sea creatures routinely utilize calcium carbonate as the basic building block from which they make the hard structures—like reefs or shells—on which their lives depend.

The harsh impact of this extra CO_2 in our oceans is depicted in the three maps of the Western Hemisphere on the far right. They show the ideal ocean conditions from a coral's point of view. The large green area in the top picture represents the optimal conditions for coral, as found in the preindustrial era. The middle picture shows the present-day conditions and depicts the shrinking of the areas optimal for corals because of the increased acidity in the oceans.

MORNING SUN SEA STAR, BRITISH COLUMBIA, CANADA

The bottom picture projects what would happen to the ocean acidity if we allow the doubling of preindustrial CO_2 levels—which will occur within 45 years unless we do something about it. As the picture shows, the optimal areas for coral would completely disappear.

Perhaps surprisingly, the decreased saturation of calcium carbonate caused by the extra CO_2 begins in the colder ocean waters near the poles, and then, as the volume of CO_2 goes up, the acidification moves from the poles toward the Equator.

The photograph to the left shows a sea star, one of the many life forms harmed by higher CO_2 levels in the ocean.

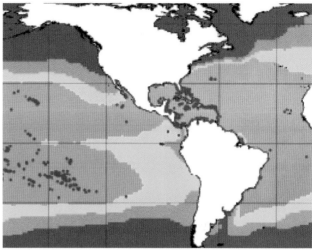

Preindustrial, 1880

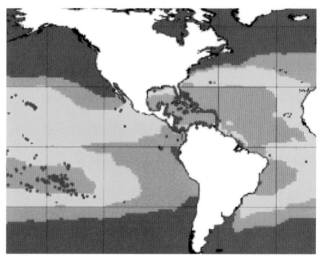

Current, 2000

OPTIMAL CORAL HABITAT
CALCIUM CARBONATE
SATURATION IN OCEAN
SURFACE WATERS

>4.0 Optimal

3.5–4.0 Adequate

3.0–3.5 Marginal

<3.0 Extremely low

SOURCE: USGCRP

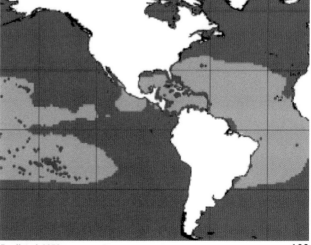

Predicted, 2050

We are changing the chemistry of our oceans in many ways, all over the world. As a result, there are many new "dead zones" devoid of ocean life. Some are caused by the appearance of algae blooms in warmer waters fed by pollution coming from human activities on the shore.

Many of these algae blooms have grown to spectacular and totally unprecedented levels in several places. In the Baltic Sea, for example, many resorts had to be closed in the summer of 2005 as a result of algae.

Florida's red tide represents a similar phenomenon.

ALGAE BLOOMING IN THE BALTIC
SEA, GOTLAND, SWEDEN, 2005

ALGAE BLOOMING NEAR THE
COAST, GOTLAND, SWEDEN, 2005

ALGAE BLOOMING IN THE BALTIC
SEA, GOTLAND, SWEDEN, 2005

Algae is just one of the disease vectors that have been increasing in range because of global warming. And when these vectors—whether algae, mosquitoes, ticks, or other germ-carrying life forms—start to show up in new areas and cover a wider range, they are more likely to interact with people, and the diseases they carry become more serious threats.

In general, the relationship between the human species and the microbial world of germs and viruses is less threatening when there are colder winters, colder nights, more stability in climate patterns, and fewer disruptions. In addition, the threat we face from microbes is reduced when the rich biodiversity of areas such as tropical rain forests—where the largest percentage of species on the planet is located—is protected from destruction and human encroachment.

VECTORS FOR EMERGING INFECTIOUS DISEASES

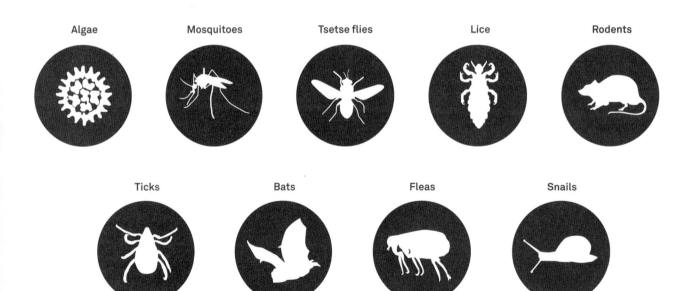

Algae   Mosquitoes   Tsetse flies   Lice   Rodents

Ticks   Bats   Fleas   Snails

Global warming pushes all of these boundaries in the wrong direction, thereby increasing human vulnerability to new and unfamiliar diseases, as well as new strains of diseases once under control.

To cite one important example of this phenomenon, mosquitoes are profoundly affected by global warming. There are cities that were originally located just above the mosquito line, which used to mark the altitude above which mosquitoes would not venture. Nairobi, Kenya, and Harare, Zimbabwe, are two such cities. Now, with global warming, the mosquitoes are climbing to higher altitudes.

MOSQUITOES MOVE TO HIGHER ELEVATIONS

TODAY
Increased warmth has caused some mosquitoes and mosquito-borne diseases to migrate to higher altitudes.

BEFORE 1970
Cold temperatures caused freezing at high elevations and limited mosquitoes and mosquito-borne diseases to low altitudes.

Some 30 so-called new diseases have emerged over the last 25 to 30 years. And some old diseases that had been under control are now surging again.

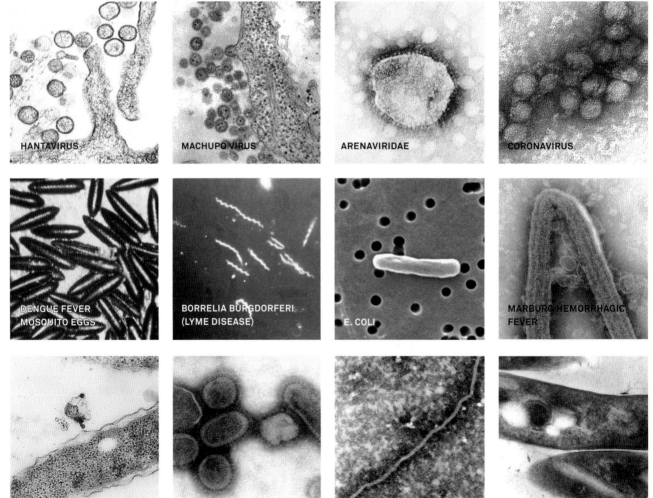

HANTAVIRUS

MACHUPO VIRUS

ARENAVIRIDAE

CORONAVIRUS

DENGUE FEVER MOSQUITO EGGS

BORRELIA BURGDORFERI (LYME DISEASE)

E. COLI

MARBURG HEMORRHAGIC FEVER

LEGIONNAIRES DISEASE

INFLUENZA VIRUS

NIPAHVIRUS

TUBERCULOSIS

One example is West Nile virus, which entered the United States on the eastern shore of Maryland in 1999 and within two years crossed the Mississippi. Two years after that, West Nile spread all the way across the continent.

SPREAD OF WEST NILE VIRUS IN THE UNITED STATES

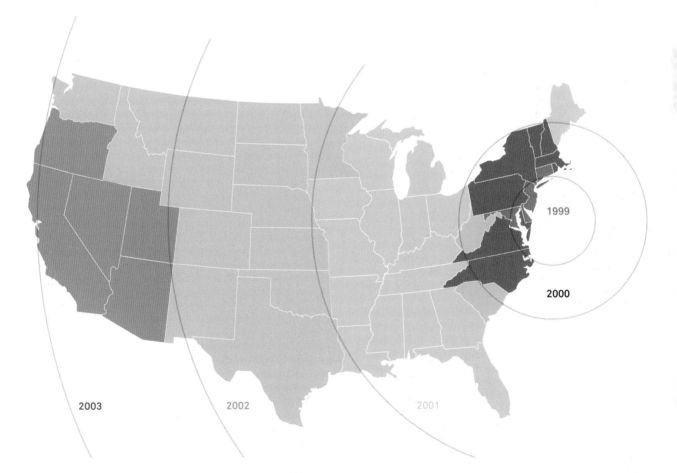

SOURCE: COMPILED FROM CDC, HEALTH CANADA, USGS, AND PROMED-MAIL SOURCES AS OF MAY 14, 2003

The second canary in the coal mine — along with the Arctic—is Antarctica, the largest mass of ice on the planet by far.

Antarctica is the closest thing to another planet we can experience on this one. It is surreal—completely and unremittingly white in every direction, so vast and so cold—much colder than the Arctic. The enormity of all that snow masks a surprising fact: Antarctica is actually a desert. It meets the technical definition in that it receives less than one inch of precipitation per year. Think about it—an icy desert, a freeze-dried oxymoron.

Antarctica is neutral territory. It is governed by an international treaty that prevents any territorial claims or military activity and reserves the entire continent for peaceful scientific endeavors, which are pursued by more than a dozen countries. The United States has the largest presence in Antarctica, under the auspices of the National Science Foundation, and also operates the Amundsen-Scott South Pole Station.

The principal U.S. base of operations, Ross Island, sits on the edge of the continent, where it can be supplied by ships during the summer. The island base is bound by permanent sea ice to the nearby mainland at McMurdo Sound, due south of New Zealand, beyond the roughest seas on the planet. Most visitors fly in on specially configured planes that land on an ice runway open for only part of the year. Ominiously, Ross Island experienced its first recorded rain just a few years ago.

Significant numbers of penguins, seals, and birds hug the edge of Antarctica and manage to find food in the ocean. But beyond the outer edge of the continent, there are absolutely no signs of life— other than the small groups of scientists who usually do not venture too far or too long from their heated enclosures.

ANTARCTICA

There may not be any real canaries in Antarctica, but there are birds—the most famous of which are these Emperor penguins, who starred in the 2005 documentary film *March of the Penguins.*

But one fact left out of the movie is that the population of Emperors has declined by an estimated 70% over the past 50 years—and scientists suspect that the principal reason is global warming.

MARCH OF THE PENGUINS

Audiences of the popular documentary *March of the Penguins* could be forgiven for thinking that the biggest challenge facing Antarctica's Emperor penguins is their icy cold habitat. In fact, a significant threat to these denizens of the southernmost continent is that their home won't be icy enough for very much longer. Scientists studying Emperor penguins at the colony featured in the film found that their numbers have dropped by 70% since the 1960s. The likely culprit: global climate change.

In the 1970s, increasingly warm temperatures, in both the air and the ocean, descended on the penguins' Antarctic home. The Southern Ocean experiences natural shifts in weather from one decade to the next, but this warm spell has continued practically unabated. Warmer temperatures and stronger winds produce thinner sea ice, the frozen ocean water on which the penguins nest. The weakened ice is more likely to break apart and drift out to sea, taking the penguins' eggs and chicks with it. Emperor penguins are the only species of bird that can survive exclusively in or on the ocean—without ever touching land. But for the sea ice to be stable enough to nest on, it must be attached to land.

Scientists believe global warming is responsible for the rising temperatures and changes to sea ice, though they can't be certain. Sea ice has decreased only in certain parts of Antarctica, but the frozen freshwater that covers most of the land mass—called "land ice"—is thinning across the whole continent. A recent NASA study using satellite mapping

Still from *March of the Penguins.*
© Jérôme Maison / Bonne Pioche. A film by Luc Jacquet. Produced by Bonne Pioche productions.

technology found that Antarctica is losing land ice at a rate of 31 billion tons of water per year. The Emperor penguins, like other animals who rely on sea ice to breed and hunt for food, are feeling the impacts first.

EMPEROR PENGUIN FAMILY,
WEDDELL SEA, ANTARCTICA

John Mercer, a scientist whose work I first saw when I was investigating global warming as a member of Congress, said in 1978: "One of the warning signs that a dangerous warming trend is underway in Antarctica will be the break-up of ice shelves on both coasts of the Antarctic Peninsula, starting with the northernmost and extending gradually southward."

At right is the Antarctic Peninsula. Each orange splotch represents an ice shelf the size of Rhode Island or larger that has broken up since Mercer issued his warning.

The red splotch marked 2002 is the Larsen-B, also portrayed in the photograph to the left—which illustrates the massive size of the ice shelves, rising roughly 700 feet above the ocean surface.

THE LARSEN-B ICE SHELF, ANTARCTICA

DISAPPEARING ANTARCTIC PENINSULA ICE SHELF

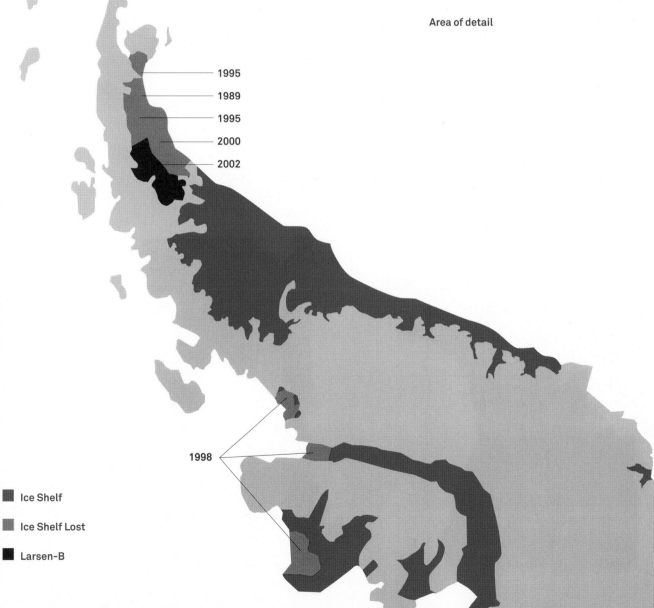

ANTARCTICA

Area of detail

1995
1989
1995
2000
2002

1998

■ Ice Shelf

■ Ice Shelf Lost

■ Larsen-B

The Larsen-B ice shelf, as photographed below, was about 150 miles long and 30 miles wide.

When you look at the black pools on top of it, it seems as if you're looking through the ice to the ocean beneath. But that's an illusion. Actually, those are pools of melting water collecting on top of the shelf.

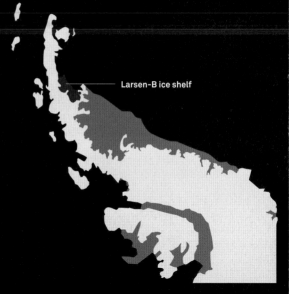

Larsen-B ice shelf

SOURCE: J. KAISER, *SCIENCE*, 2002

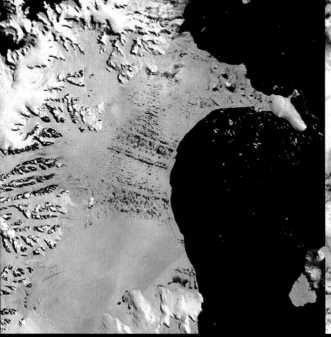

SATELLITE IMAGE OF
THE LARSEN-B ICE SHELF,
JANUARY 31, 2002

FEBRUARY 17, 2002

Scientists thought this ice shelf would be stable for at least another century—even with global warming. But starting on January 31, 2002, within 35 days, it completely broke up. Indeed, most of it disappeared over the course of two of those days. Scientists were absolutely astonished. They couldn't figure out how in the world this had happened so rapidly. So they went back to assess why their estimates were so off.

They found that they had made an incorrect assumption about those melting pools of water on top of the ice mass. They had thought that the meltwater sank back into the ice and refroze. Instead, as they now know, the water keeps sinking straight down and makes the ice mass look like Swiss cheese.

SOURCE: MODIS IMAGES COURTESY OF NASA'S TERRA SATELLITE

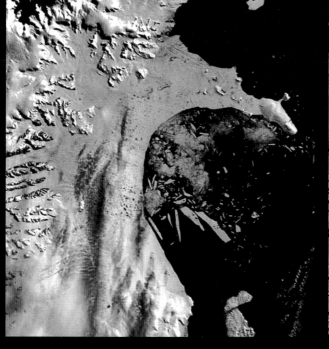

FEBRUARY 23, 2002

MARCH 5, 2002

Once the sea-based ice shelf was gone, the land-based ice behind it that was being held back began to shift and fall into the sea. This, too, was unexpected and carries important implications because ice—whether in the form of a mountain glacier or a land-based ice shelf in Antarctica or Greenland—raises the sea level when it melts or falls into the sea.

This is one of the reasons sea levels have been rising worldwide, and will continue to go up if global warming is not quickly checked.

CALVING FRONT OF YAHTSE GLACIER, WRANGELL-ST. ELIAS NATIONAL PARK, AK, 1995

Many residents of low-lying Pacific Island nations have already had to evacuate their homes because of rising seas.

HIGH TIDE IN FUNAFUTI, TUVALU, POLYNESIA

The Thames River, which flows through London, is a tidal river. In recent decades, higher sea levels began to cause more damage during storm surges, so a quarter of a century ago, the city built these barricades that can be closed for protection.

The graph below shows how frequently London has had to use these barriers in recent years. The data for years prior to the construction of the barriers represent the city's projections of what the number of closures would have been in those earlier years. The resulting pattern is similar to many others that measure the increasing impact of global warming worldwide.

But further sea level rises could be many times larger and more rapid depending on what happens in Antarctica and Greenland—and on choices we make or do not make—now concerning global warming.

ANNUAL CLOSURES OF THE THAMES BARRIERS

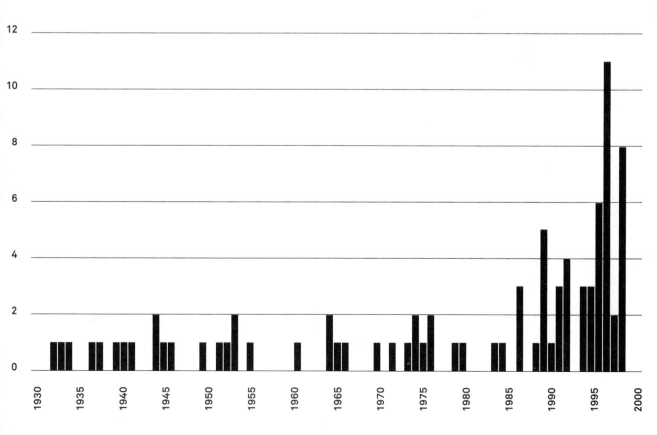

SOURCE: UK ENVIRONMENT AGENCY

Now consider the much larger areas of ice in Antarctica and Greenland that are at risk.

The East Antarctic ice shelf is the largest ice mass on the planet and had been thought to be still increasing in size. However, two new studies in 2006 showed first that the overall volumes of ice in East Antarctica now appear to be declining, and that 85 percent of the glaciers there appear to be accelerating their flow toward the sea. Second, it showed that air temperatures measured high above this mass of ice appear to have warmed more rapidly than air temperatures anywhere else in the world. This finding was actually a surprise, and scientists have not yet been able to explain why it is occurring.

East Antarctica is still considered far more stable over long periods of time than the West Antarctic ice shelf, which is propped up against the tops of islands. This peculiar geology is important for two reasons: first, its weight is resting on land and therefore its mass has not displaced seawater as floating ice would. So if it melted or slipped off its island moorings into the sea, it would raise sea levels worldwide by 20 feet. Second, the ocean flows underneath large sections of this ice shelf, and as the ocean has warmed, scientists have documented significant and alarming structural changes on the underside of the ice shelf.

Interestingly, the West Antarctic ice shelf is virtually identical in size and mass to the Greenland ice dome, which also would raise sea levels worldwide by 20 feet if it melted or broke up and slipped into the sea.

WEST ANTARCTIC
ICE SHELF

EAST ANTARCTICA

GREENLAND

These photographs from Greenland illustrate some of the dramatic changes now happening on the ice there. I flew over Greenland in 2005 and saw for myself the pools of meltwater covering large expanses on top of the ice. The picture at left, below, taken recently by a friend of mine, Dr. Jim McCarthy of Harvard University, shows one such area. These pools have always been known to occur, but the difference now is that there are many more of them covering a far larger area of the ice. This is significant because, as Dr. McCarthy points out, they are exactly the same kind of meltwater pools that he and other scientists observed on top of the Larsen-B ice shelf in the period before its sudden and shocking disappearance.

In Greenland, as in the Antarctic Peninsula, this meltwater is now believed to keep sinking all the way down to the bottom, cutting deep crevasses and vertical tunnels that scientists call "moulins."

When the water reaches the bottom of the ice, it lubricates the surface of the bedrock and destabilizes the ice mass, raising fears that the ice mass will slide more quickly toward the ocean.

Below is an illustration of how meltwater tunnels downward through crevasses and moulins in the ice on Greenland. At right is an actual moulin—a massive torrent of fresh meltwater tunneling straight down to the bedrock below the ice. Notice, for scale, the scientists at the top of the picture.

MELTWATER POOLS, GREENLAND, 2005

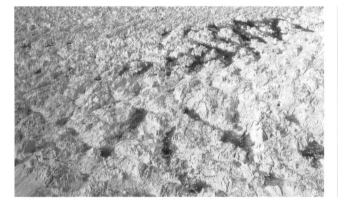

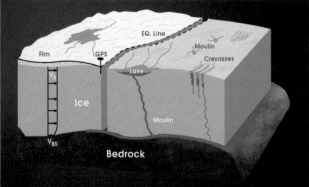

STREAM OF MELTWATER
CASCADING OFF THE ARCTIC ICE
SHEET, GREENLAND, 2005

To some extent, there has always been seasonal melting, and moulins have formed in the past. But those formations were nothing like what is happening now. In recent years, the melting has accelerated dangerously.

In 1992 scientists measured this amount of melting in Greenland as indicated by the red areas on the map.

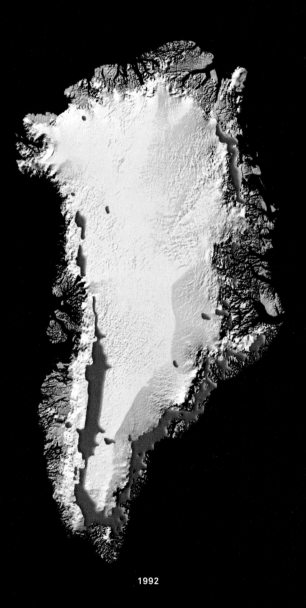

1992

Ten years later, in 2002, the melting was much worse.

And in 2005 it accelerated dramatically yet again.

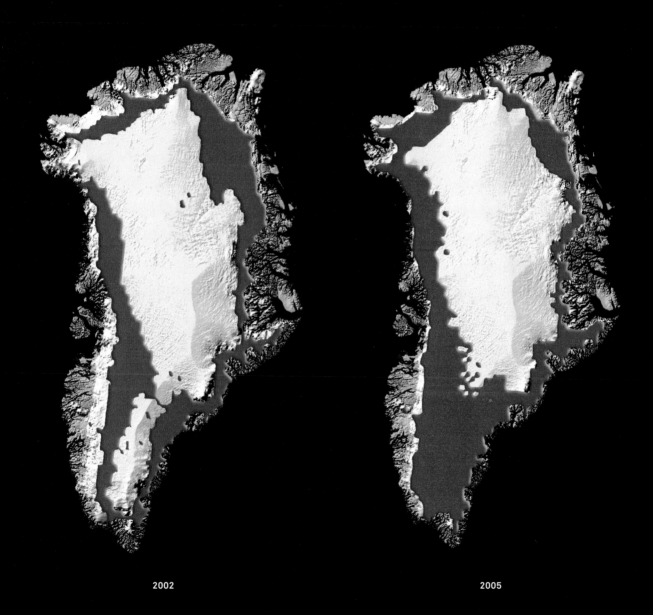

2002

2005

SOURCE: ©2005 ACIA

If Greenland melted or broke up and slipped into the sea—or if half of Greenland and half of Antarctica melted or broke up and slipped into the sea, sea levels worldwide would increase by between 18 and 20 feet.

Tony Blair's advisor, David King, is among the scientists who have been warning about the potential consequences of large changes in these ice shelves.
At a 2004 conference in Berlin, he said:

THE MAPS OF THE WO TO BE REDRAWN.

SIR DAVID KING, U.K. SCIENCE ADVISOR

RLD WILL HAVE

This is what would happen to Florida.

This is what would happen to San Francisco Bay.

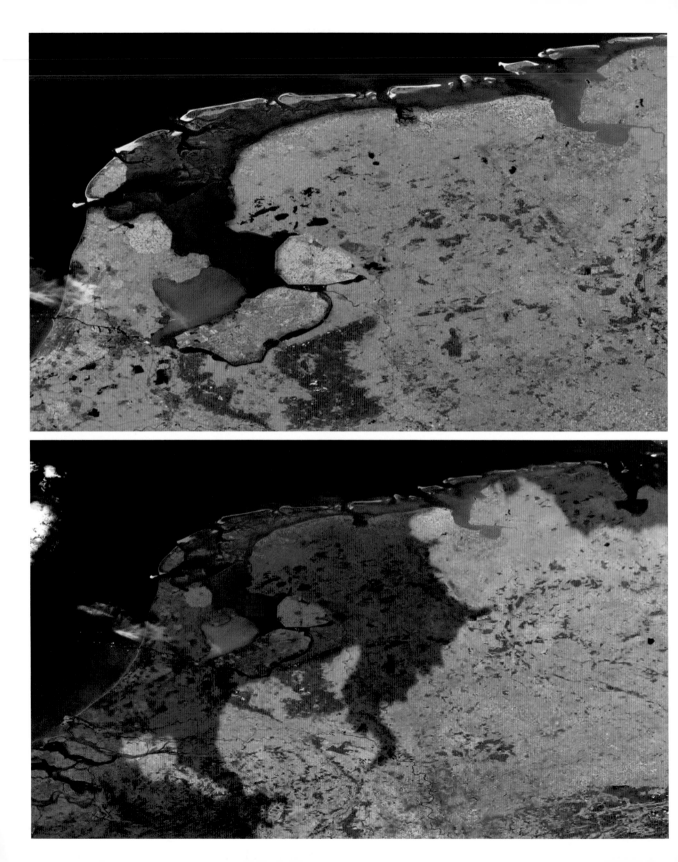

This is what would happen to the Netherlands, one of the low countries where rising sea levels would be absolutely devastating.

But the Dutch, who have long experience in dealing with the sea, have already launched a competition among architects to design floating homes. One of them is pictured at the right.

FLOATING HOUSES, AMSTERDAM, THE NETHERLANDS, 2000

Here's what would happen to Beijing and the surrounding area. More than 20 million people would have to be evacuated.

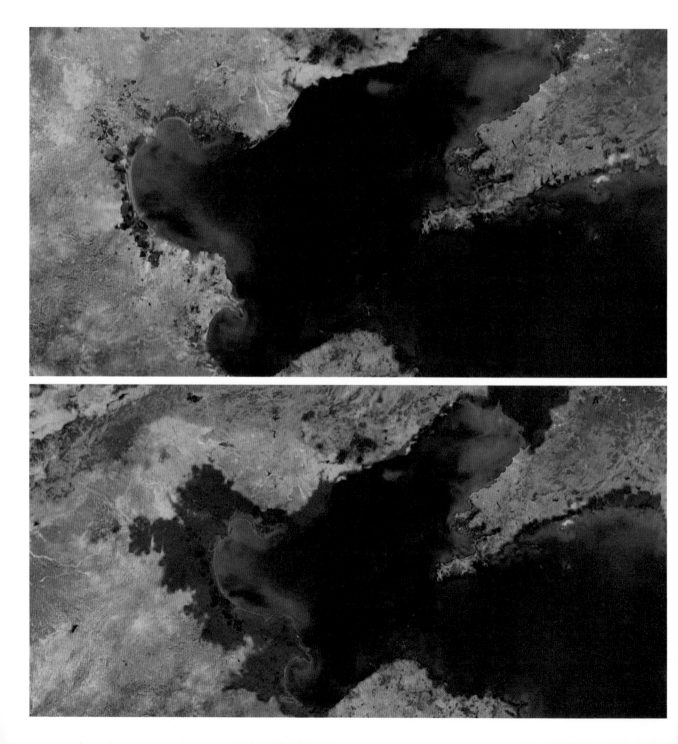

In Shanghai and the surrounding area,
more than 40 million people would be
forced to move.

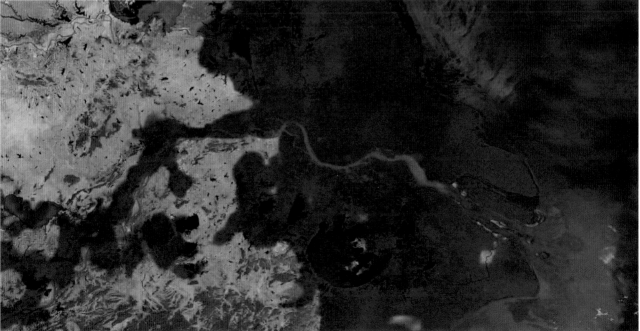

In Calcutta and Bangladesh, 60 million people would be displaced.

In Manhattan, the World Trade Center Memorial is intended to be, among other things, an expression of the determination of the United States never to allow such harm to befall our country again.

But the picture at right shows what would happen to Manhattan if sea levels rose 20 feet worldwide. The site of the World Trade Center Memorial would be underwater.

WORLD TRADE
CENTER MEMORIAL

Is it possible that we should prepare for other serious threats in addition to terrorism? Maybe it's time to focus on other dangers as well.

WORLD TRADE
CENTER MEMORIAL

Serving for the Public Good

— ◆ —

**American constitutional democracy
still has the potential to confer on the
average citizen the dignity and the
majesty of self-governance.**

My father was a hero to me. I looked up to him and I wanted to be like him. For the same reason that so many boys have chosen to follow in their father's footsteps, I thought at a very young age that public service might well be the path I should choose for my career—just like my Dad.

Albert Gore Sr., had served in the U.S. Congress for 10 years by the time I was born, in 1948. He was elected to the Senate when I was four years old and didn't leave it until I had graduated from college and was serving with the U.S. Army in Vietnam. He ultimately spent a total of 32 years in the House and Senate combined. He was a strong and courageous man with vision and integrity. As a young boy, I thought: why *wouldn't* I want to be like him?

But as I grew older, two things happened. First, I watched as my father was defeated for reelection to the U.S. Senate, in 1970, mainly because of his courageous opposition to the Vietnam War, his support for school desegregation and voting rights, and his insistence on upholding constitutional principles against the onslaught being carried out by the Nixon-Agnew administration. Second, I saw politics in the United States changing from the way I had experienced it when I was younger. Negative TV ads, for example, brought an ugly and mean new tone into politics; now it no longer had the same appeal for me. After serving in Vietnam as an army journalist, I came back home feeling utterly disconnected from my early childhood interest in politics; I thought politics would be the very last thing I did with my life. So I went to work for *The Nashville Tennessean* as a newspaper reporter.

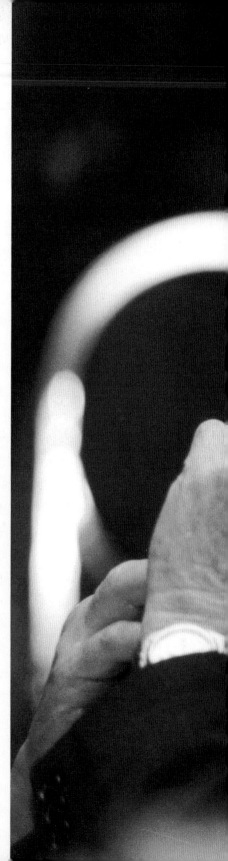

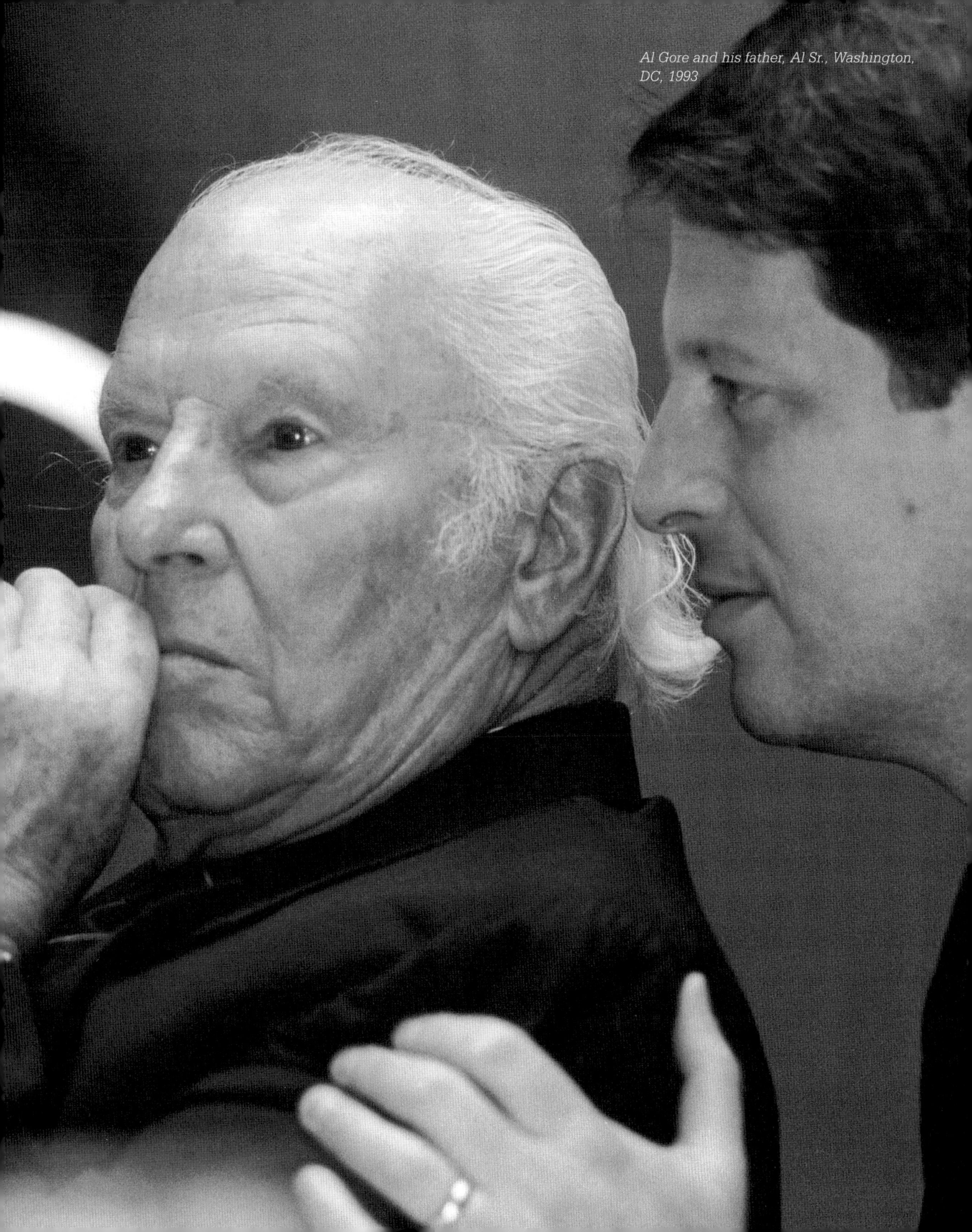

Al Gore and his father, Al Sr., Washington, DC, 1993

**What struck me most was the palpable thrill I felt at making democracy
work the way I felt it was supposed to: listening to people, discussing
their ideas with them, and then trying to make practical sense
of those ideas within the context of the legislative process.**

But after spending five years covering public affairs as a journalist, I began to rekindle my interest in the democratic process. Seeing politics up close from that different and detached perspective, I slowly but surely came back to it on my own terms. Then, in 1976, when the congressman from my home district surprised everyone by announcing his decision to retire, I jumped into the race and narrowly won election to Tennessee's fourth congressional district.

In those days, winning the Democratic primary in Middle Tennessee was tantamount to winning the general election,

because there were so few Republican voters there that they didn't even nominate a candidate for the general election. So, following the August primary, I began to chart my course in Congress, even though I would not be sworn into office until the following January. My first act was to travel to the Oak Ridge National Laboratory to spend several days steeping myself in the latest research on energy and the environment. Even back then these were at the top of my list of priorities. Then I undertook a series of town meetings (I called them "open meetings," and I had lots of them) throughout the

25 counties I had just been selected to represent.

What struck me most was the palpable thrill I felt at making democracy work the way I felt it was supposed to: listening to people, discussing their ideas with them, and then trying to make practical sense of those ideas within the context of the legislative process. This exhilaration was new to me. While it connected with my childhood impulse toward public service, it went beyond that. It was much more deeply felt. It was intense and real. It felt good and I loved it.

There are plenty of reasons to feel cynical about the way democracy works in America today and about some of the men and women who are candidates and elected officials. I do understand why so many people are discouraged about the performance of American government, especially in recent years.

But despite the negatives, there is something powerful and resilient we must never lose sight of: American constitutional democracy still has the potential to confer on the average citizen the dignity and majesty of self-governance. It is still, in Churchill's well-known phrase, "the worst form of government except all

*Al, with his parents, celebrating his
father's reelection to the U.S. Senate,
Nashville, TN, 1958*

Al Gore, with wife and daughter, announcing his first campaign for Congress, Carthage, TN, 1976

the others that have been tried." When it works the way our founders intended, the very act of self-governing can produce an indescribable feeling of goodness and harmony that no cynic will ever be able to diminish.

The most important thing I learned about American democracy—I learned first from watching my father, from my interaction with the people in my congressional district, and throughout my service as a senator and vice president—is that the spirit of freedom that motivated Thomas Paine, Patrick Henry, our founders, and the true patriots of every genera-

tion since, is always present, just waiting for the right spark to ignite it.

Since I left the White House in 2001, I have also learned that there are many forms of public service other than running for office and serving as an elected official. I've always known this, of course, but I've come to personally appreciate the satisfaction that can be found as a private citizen in trying to make our democracy work better. Speaking out about the issues confronting our nation, and trying to describe the challenges we face and the solutions available, is a form of public service our founders described

as essential to our democracy's survival. James Madison wrote that "a well-informed citizenry" is the bedrock of the American constitutional system. I guess I didn't expect to enjoy serving in the capacity of "citizen" as much as I do.

But when I get as close as I'm capable of to laying out an important truth as I see it, and then helping to connect that truth to a course of action, I actually do experience the same feeling I had back in 1976 when I was making the rounds of my new congressional district in Middle Tennessee. I'm really chasing that feeling in writing this book.

We are witnessing an unprecedented and massive collision between our civilization and the Earth.

REFUSE DUMP IN MEXICO CITY, MEXICO, 1996

The fundamental relationship between our civilization and the ecological system of the Earth has been utterly and radically transformed due to the powerful convergence of three factors.

The first is the population explosion, which in many ways is a success story in that death rates and birth rates are going down everywhere in the world, and families, on average, are getting smaller. But even though these hoped-for developments have been taking place more rapidly than anyone would have anticipated a few decades ago, the momentum in world population has built up so powerfully that the "explosion" is still taking place and continues to transform our relationship to the planet.

If you look at population growth in the context of history, it is obvious that the last 200 years represent a complete break with the pattern that prevailed for most of the millennia that humans have walked on the Earth. From the time when scientists say our species first appeared 160,000 to 190,000 years ago, until the time of Jesus Christ and Julius Caesar, human population had grown to a quarter of a billion people. By the time of America's birth in 1776, it had grown to 1 billion. When the baby boom generation that I'm a part of was born at the end of World War II, the population had just crossed 2 billion. In my lifetime, I have watched it go all the way to 6.5 billion. My generation will see it rise to more than 9 billion people.

The point, as illustrated by this graph, is simple and powerful: It took more than 10,000 generations for the human population to reach 2 billion. Then it began to rocket upward from 2 billion to 9 billion in the course of a single lifetime: ours. We have a moral obligation to take into account this dramatic change in terms of the relationship between our species and the planet.

POPULATION GROWTH THROUGHOUT HISTORY

First modern humans

160,000 BC   100,000 BC   10,000 BC   7000 BC   6000 BC   5000 BC   4000 BC

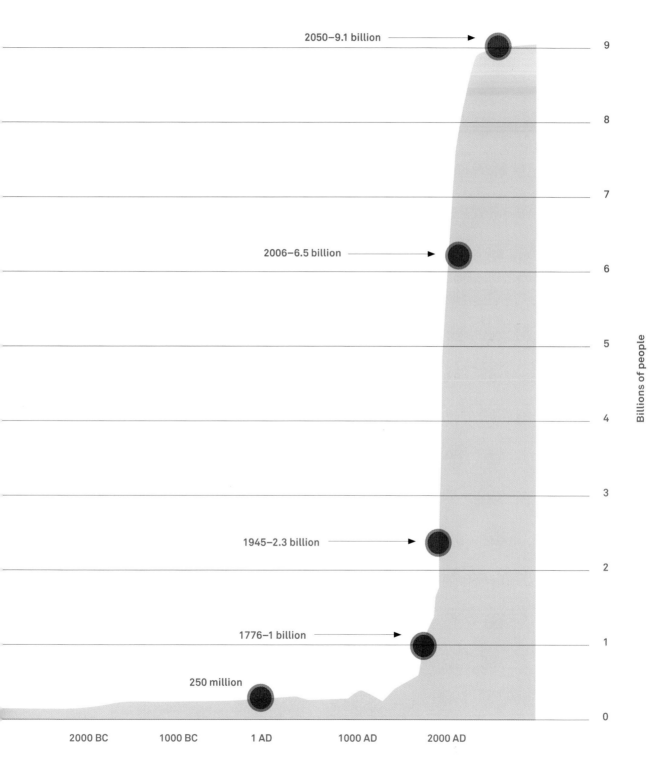

2050−9.1 billion

2006−6.5 billion

1945−2.3 billion

1776−1 billion

250 million

Billions of people

9

8

7

6

5

4

3

2

1

0

2000 BC   1000 BC   1 AD   1000 AD   2000 AD

Most of the increase in population is
in developing nations where most of the
world's poverty is concentrated.

SHINJUKU DISTRICT OF TOKYO,
JAPAN, 1996

And most of the increase is in cities.

This rapid population rise drives demand for food, water, and energy—and for all our natural resources. It puts enormous pressure on vulnerable areas like forests—particularly the rain forests of the tropics.

LOGGING, TAPAJOS NATIONAL FOREST, BRAZIL, 2004

STUMPS AND SLASH AFTER CLEARCUTTING, NEAR FORKS, WA, 1999

The way we treat forests is a political issue.

HAITI

This is the border between Haiti and the Dominican Republic. Haiti has one set of policies; the Dominican Republic another.

DOMINICAN REPUBLIC

The Amazon is suffering particular devastation. Here are two satellite images of the Rondonia region of Brazil, taken 26 years apart.

RONDONIA, BRAZIL, 1975.

RONDONIA, 2001

Much of the forest destruction comes from burning. Almost 30% of the CO_2 released into the atmosphere each year is a result of the burning of brushland for subsistence agriculture and wood fires used for cooking.

**FARM WORKER BURNING
RAIN FOREST TO CLEAR LAND
FOR RANCHING, RONDONIA,
BRAZIL, 1988**

Wildfires are becoming much more common as hotter temperatures dry out the soil and the leaves. In addition, warmer air produces more lightning. The graph below shows the steady increase in major wildfires in North and South America over the last five decades; the same pattern is found on every other continent as well.

NUMBER OF MAJOR WILDFIRES IN THE AMERICAS BY DECADE

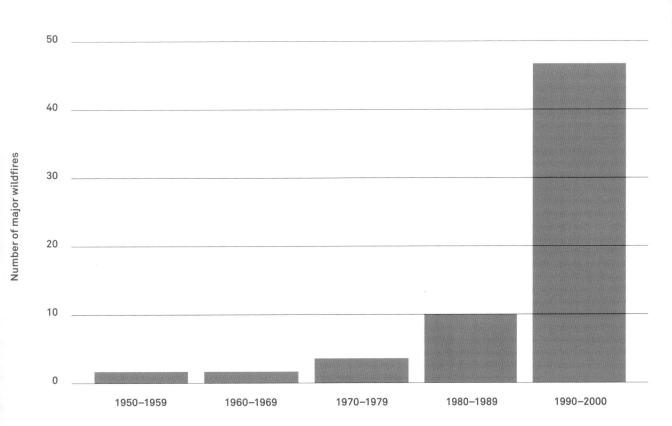

SOURCE: MILLENNIUM ECOSYSTEM ASSESSMENT

This is a six-month time-lapse image of the world at night from a U.S. Defense Department satellite. The lights of the cities are shown in white. The blue areas depict the enormous fishing fleets active at night, particularly in Asia and Patagonia. All the parts of the Earth colored red are the places where fires have been burning. Africa stands out partly because of the prevalence of wood fires for cooking. The yellow areas are the gas flares in the oil fields. The Siberian oil fields appear larger than the Persian Gulf because more of the gas in the Persian Gulf is captured now, rather than burned.

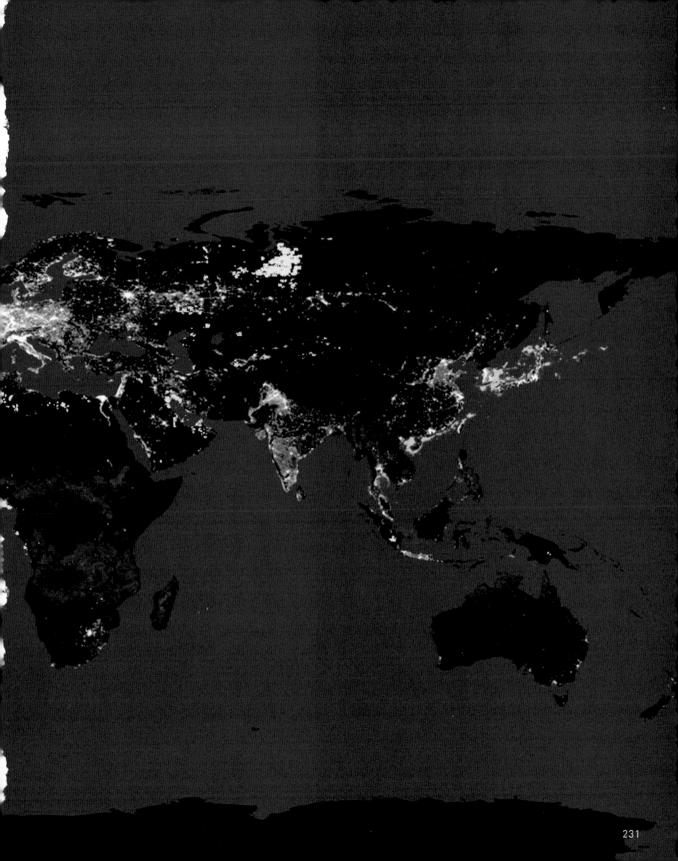

231

And that brings me to the second factor that has transformed our relationship to the Earth—the scientific and technological revolution.

New advances in science and technology have brought us tremendous improvements in areas like medicine and communications, among many others. For all the advantages we have gained from our new technologies, we have also witnessed many unanticipated side effects.

The new power we have at our disposal hasn't always been accompanied by new wisdom in the way we use it, particularly when we exercise our technologically enhanced power in the thoughtless pursuit of age-old habits, which are, after all, hard to change.

The simple formulas below are intended to illustrate what can happen when vastly more powerful technologies magnify the consequences of doing the same old thing without anticipating that there will be brand-new consequences.

$$\frac{\text{Old habits}}{\text{+}} \text{Old technology} = \text{Predictable consequences}$$

**ASTRONOMICAL RADIO OBSERVATORY,
SOCORRO, NM**

Old
habits
+
New
technology

=

Dramatically
altered
consequences

Here is an example of how these formulas work.

Warfare is an ancient habit. The consequences of warfare waged with the technology of spears and swords—or bows and arrows, or muskets and rifles—were horrible, but predictable.

In 1945, however, the brand-new technology of nuclear weapons completely altered the equation.

As a result, we have tried to reevaluate, and change, the old habit we call war. Though we have made some progress— we had a Cold War instead of an all-out nuclear war—we still have a lot of work to do.

TOP: VASE DEPICTING A HOPLITE BATTLE, C. 600 BC. BOTTOM: PAINTING OF THE BATTLE OF CHIPPEWA, 1812

TOP: PAINTING OF A CRUSADER BATTLE, C. 1250. BOTTOM: GERMAN SOLDIERS IN WORLD WAR I, 1914

TEST DETONATION OF A
NUCLEAR BOMB, NV, 1957

Similarly, we have always exploited the Earth for sustenance, utilizing relatively basic technologies for most of our existence, like plowing and irrigating and digging into the earth. But even these simple technologies have become far more powerful today.

**FARMER PLOWING A FIELD,
PATTANI, THAILAND, 1966**

FARMER TEDDING HAY,
IA, 2000

We now have a much more profound ability to transform the surface of the planet. In the same way, every human activity is now pursued with much more powerful tools, which often bring unanticipated consequences.

COPPER STRIP MINE, CANNEA, MEXICO, 1993

Irrigation has long worked wonders for humankind. But we now have the power to divert giant rivers according to our design instead of nature's.

241

When we divert too much water without regard for nature, rivers sometimes no longer reach the sea.

VIEW UPSTREAM FROM
THE HITE LOOKOUT, COLORADO
RIVER, AZ, 2002

The former Soviet Union diverted water from two mighty rivers in central Asia that fed the Aral Sea (the Amu Darya and Syr Darya). These two rivers were used for irrigating cotton fields.

When I went there some years ago, I saw a strange sight: an enormous fishing fleet marooned in the sand, with no water in sight. This picture shows part of that fleet and the canal that the fishing industry desperately tried to dig in an effort to chase the receding shoreline.

GROUNDED FISHING BOATS,
ARAL SEA, KAZAKHSTAN, 1990

The entire Aral Sea is now, essentially, gone.

The parable of the Aral Sea has a simple message: Mistakes in our dealings with Mother Nature can now have much larger, unintended consequences, because many of our new technologies confer upon us new power without automatically giving us new wisdom.

Indeed, as illustrated by this image, some of our new technologies overwhelm the human scale.

COAL-FIRED POWER STATION,
FERRYBRIDGE, ENGLAND

COMPOSITE SATELLITE VIEW OF
THE EARTH AT NIGHT, 1994–1995

Our new technologies, combined with our numbers, have made us, collectively, a force of nature.

249

And those with the most technology have the greatest moral obligation to use it wisely. And this, too, is a political issue. Policy matters.

As shown in this graphic representation of every nation's relative contribution to global warming, the United States is responsible for more greenhouse gas pollution than South America, Africa, the Middle East, Australia, Japan, and Asia— all put together.

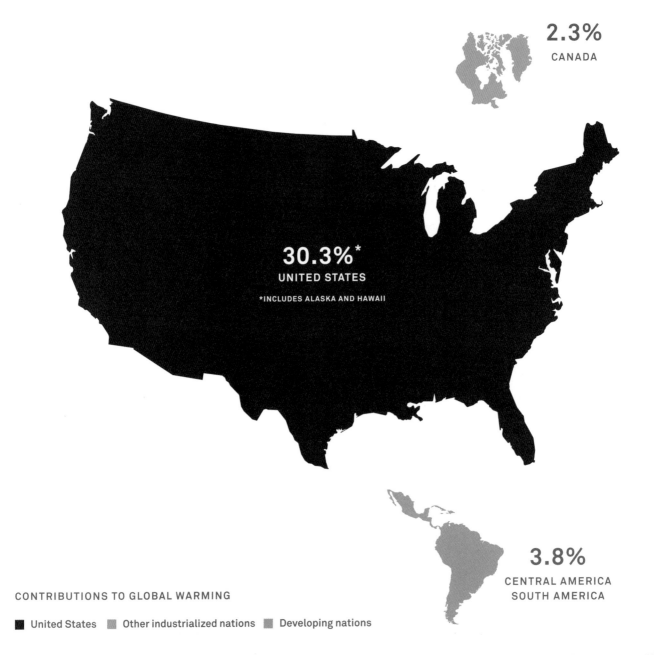

2.3%
CANADA

30.3%*
UNITED STATES
*INCLUDES ALASKA AND HAWAII

3.8%
CENTRAL AMERICA
SOUTH AMERICA

CONTRIBUTIONS TO GLOBAL WARMING

■ United States ■ Other industrialized nations ■ Developing nations

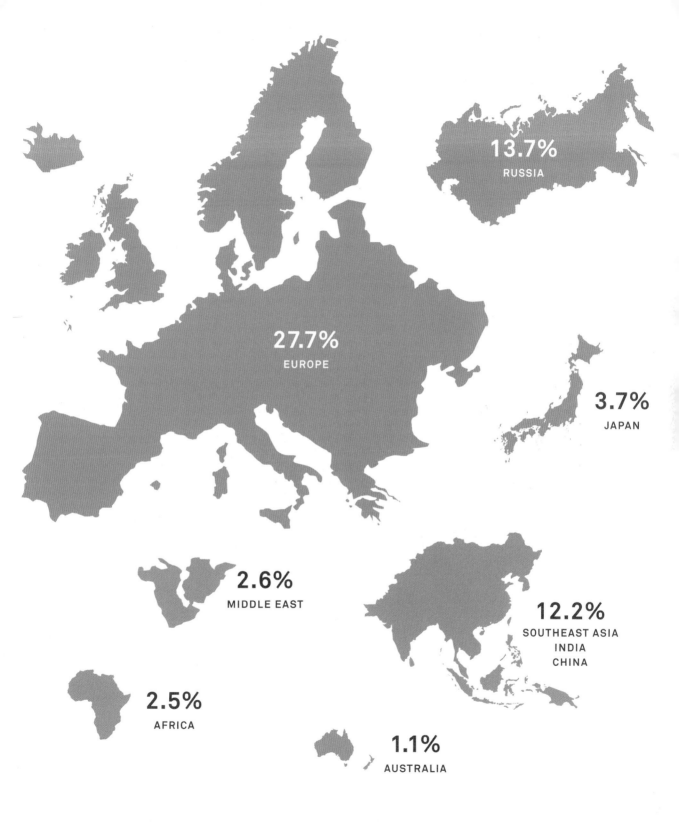

13.7%
RUSSIA

27.7%
EUROPE

3.7%
JAPAN

2.6%
MIDDLE EAST

12.2%
SOUTHEAST ASIA
INDIA
CHINA

2.5%
AFRICA

1.1%
AUSTRALIA

If you compare the per capita carbon emissions in China, India, Africa, Japan, the EU, and Russia to those in the United States, it is obvious, as the chart at top right shows, that we are way, way above everyone else.

Of course, population size must be factored into the calculation. And when it is, as illustrated in the graph at bottom right, China's role is seen as a larger (and growing) factor. The same is true of Europe. But the United States is still far above the others.

THE CARBON EXCHANGE MARKET

When acid rain was falling on parts of the United States—particularly in the Northeast—back in the 1980s, an innovative program helped to clean up the polluted precipitation. With bipartisan support, the Congress put in place a system for buying and selling emissions of sulfur dioxide (SO_2), the main culprit behind acid rain. Called a cap-and-trade system, it used the power of market forces to help drastically reduce SO_2 emissions, while allowing pioneering companies to profit from environmental stewardship.

A similar approach can speed up the reduction of CO_2 emissions. The European Union has adopted this U.S. innovation and is making it work effectively. Here at home, while the Congress has not yet passed a federal cap-and-trade system for carbon dioxide emissions, there is an effective private-sector carbon market that is already up and running. The Chicago Climate Exchange (CCX) is a fledgling market based on the simple premise that reducing carbon emissions is valuable—not just as an idealistic goal, but literally worth money.

With leading businesses like Ford, Rolls-Royce, IBM, and Motorola signing on to this experiment, it's clear that some business leaders have seen the need to get a grip on global climate change and are thinking outside the box about ways to do it. One goal of the CCX is to figure out how best to run a carbon market so

that if and when our government rolls out a national cap-and-trade program (as many expect it eventually will), the kinks will already be worked out.

At this point, CCX members join up voluntarily, pledging to reduce their emissions of greenhouse gases (six are targeted). Once each member's emissions are converted into tradable credits, the exchange works like any financial market. If participants reduce their emissions below their target, they can sell their carbon credits on the exchange for a profit. If they fail to reduce their emissions, they must buy credits from others.

The value of the carbon depends on how many companies are buying it rather than selling it. For now, because more companies are selling emissions credits (because they have surpassed their reduction targets), the price of carbon is low. In Europe, though, where a carbon market is much more advanced, it trades at a much higher price. The carbon prices on the European exchanges are higher precisely because the allowances for carbon emissions are mandated by government.

In other parts of the world, there is also movement in this direction. In Canada the Montreal Commodity Exchange and the Mumbai Commodity Exchange in India, to name two, are in the process of launching carbon trading facilities. These are encouraging developments that move

the world closer to the ultimate vision of a closed global carbon market linking up the capabilities of all exchanges into a single system.

Here in the United States there has been progress at the state level in establishing mandated emissions-trading, including the Regional Greenhouse Gas Initiative in the Northeastern states and legislation is pending in California. For now, however, the private CCX initiative is where the real action is taking place.

For many companies that have already signed on, participation is a way to gain early experience with this sort of market. It is also a good incentive to embark on emissions reduction projects. For DuPont, one of the CCX's founding partners, a key benefit is the chance to shape rules and procedures for the trading system. During its participation, the company has been making changes to reduce its emissions, by increasing the energy efficiency of its facilities and decreasing its greenhouse-gas emissions from manufacturing.

All types of organizations can join the CCX. The members now include NGOs such as the World Resources Institute, municipalities such as the City of Oakland, California, and universities such as The University of Oklahoma.

The CCX is leading the way toward a future in which reducing greenhouse gases could bring not only environmental rewards, but financial ones too.

CARBON EMISSIONS PER PERSON

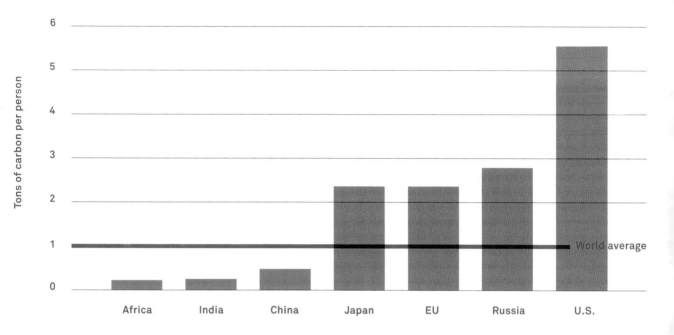

CARBON EMISSIONS PER COUNTRY/REGION

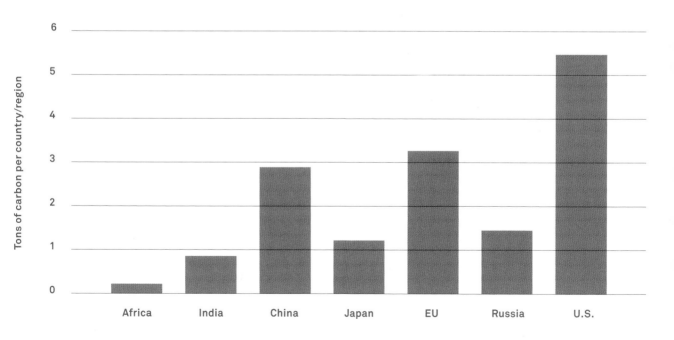

SOURCE: WORLD RESOURCES INSTITUTE
UNDERLYING DATA SOURCE: U.S. DEPARTMENT OF ENERGY, ENERGY INFORMATION ADMINISTRATION, INTERNATIONAL ENERGY ANNUAL 1999
NOTE: SHOWS CARBON EMISSIONS ASSOCIATED WITH FOSSIL FUEL COMBUSTION

The third and final factor causing the collision between humankind and nature is both the subtlest and most important: our fundamental way of thinking about the climate crisis.

And the first problem in the way we think about the climate crisis is that it seems easier not to think about it at all. One reason it doesn't consistently demand our attention can be illustrated by the classic story about an old science experiment involving a frog that jumps into a pot of boiling water and immediately jumps out again because it instantly recognizes the danger. The same frog, finding itself in a pot of lukewarm water that is being slowly brought to a boil, will simply stay in the water—in spite of the danger—until it is...rescued.

(I used to recount this story about the frog with a different ending to the last sentence above: "until the frog is boiled." But after dozens of slide shows were followed by at least one anguished listener coming up to me and expressing concern for the fate of the frog, I finally learned the importance of rescuing the frog.)

But of course the larger point of the story is that our collective "nervous system,"

DENIAL AIN'T JUST A

MARK TWAIN

through which we recognize an impending danger to our survival, is similar to the frog's. If we experience a significant change in our circumstances gradually and slowly, we are capable of sitting still and failing to recognize the seriousness of what is happening to us until it's too late. Sometimes, like the frog, we only react to a sudden jolt, a dramatic and speedy change in our circumstances that sets off our alarm bells.

Global warming may seem gradual in the context of a single lifetime, but in the context of the Earth's history, it is actually happening with lightning speed. Its pace is now accelerating so rapidly that even in our own lifetimes, we are beginning to see the telltale bubbles of a boiling pot.

We are, of course, also different from the frog. We don't have to wait for the boiling point in order to understand the danger we're in—and we do have the ability to rescue ourselves.

RIVER IN EGYPT.

My Sister

——— ❖ ———

How do I describe my sister? She was luminous. Charismatic. Gutsy. Astute. Funny. Incredibly smart. And kind.

When I was a child, despite our 10-year age difference, she was my playmate; she was also my protector. There were only the two of us kids, and we were our own team. Whether we were running through the halls of the Fairfax Hotel or diving into the Caney Fork River in Carthage, Tennessee, she knew me like no one else when I was growing up.

We both loved Center Hill Lake near our farm in Carthage. She taught me to water-ski on it. We often took canoes out on the river and had lazy conversations while we paddled. She really liked duck hunting, and got a kick out of the fact that people thought it was unusual for a young woman to duck hunt. She enjoyed breaking new ground.

Later, in her early twenties, Nancy was one of the first two volunteers for the Peace Corps, working for Bill Moyers and Sargent Shriver to open the Washington office and help get the whole thing started. I was always proud of her instinct to find work that directly improved people's lives.

She was also my constant booster, moving from her home in Mississippi during my first campaign for Congress to spend months at a time in the toughest county in the district. I credit her, in large part, with my victory in that race. She was terrific at convincing people to vote for her little brother. She was ferocious in defending me, tireless in promoting me, and more candid than anyone else could be in taking me to task when I needed it.

Nancy possessed a special grace, but also a rebellious streak. Her cigarette smoking was just one example. She started when she was 13 years old and never stopped. She tried to quit, but cigarettes really had a hold on her. The scientists have since learned that the nicotine in tobacco can be more addictive than heroin. And their studies show that kids who start smoking in their early teens or before have by far the hardest time quitting.

During the 1960s, even after the Surgeon General's report made it abundantly clear that smoking can cause lung cancer, the tobacco companies were working overtime to encourage Americans not to believe the science—to create doubts about whether there was any real cause for concern. And a lot of people who might otherwise have fully absorbed the terrible truth about smoking and health were tempted to take it less seriously than they should have. After all, if there were still such serious doubts, then maybe the jury was still out. Maybe the science wasn't definitive. So for almost 40 years after the landmark Surgeon General's report linking smoking to lung cancer, emphysema, and other diseases in the United States, more Americans continued to die from smoking-related causes than were killed during World War II.

The clever and deceitful approach the tobacco companies used to confuse people about what the science really demonstrated added up to a model for the campaign that many oil and coal companies are using today to confuse people about what the science of global warming is really telling us. They exaggerate minor uncertainties in order to pretend that the big conclusions are not a matter of consensus.

To this day, if you ask doctors and scientists to describe in intricate detail

Nancy Gore Hunger, Nashville, TN, 1964

the exact process by which smoking cigarettes leads to lung cancer, they will give you an overall picture and tell you they know for sure that there is a deadly causative link. But if you push them hard enough on some of the minor details, you will soon reach a point at which they will have to say, "Well, we don't know exactly how that particular relationship works."

Yet the fact that some important details are still to be fully understood in no way changes the reality of the problem. It's cynical and wrong to use such misleading caveats to convince people that the connection between cigarettes and cancer is a big lie. And just as it was immoral for the tobacco companies to use that tactic in the decades following the mid-1960s, it's equally immoral now for the oil and coal companies to do the same thing where global warming is concerned.

Nancy was beautiful, vibrant, and strong. But lung cancer proved too cruel an adversary. When I first got news of her initial diagnosis in 1983, I immediately visited the National Institutes of Health and spoke with leading cancer specialists about the particular form of lung cancer Nancy had and tried to learn from them how it could best be treated. Some friends have suggested that that was a

defense mechanism and that the policy wonk in me took refuge in facts and data. The truth is, I just wanted to save her.

In fact, I did what millions of Americans facing the loss of a loved one do every single day in this country: I scoured medicine for a miracle. Today, thankfully, an increasing number of cancer patients are being cured, although lung cancer remains one of the toughest types to beat. And Nancy's illness struck 20 years ago, when there wasn't the same medical arsenal there is now.

The surgeons removed one of Nancy's lungs and part of the other one. She then had to wait for a period of months to see whether that measure had been successful. It turned out not to have been.

Believe me, lung cancer is one of the ways you don't want to die. Often, the victim actually drowns from the constant build-up of fluid in the lungs once the pathology overwhelms the body's natural healing processes. The suffering can be unspeakable.

I was in the midst of my first campaign for the U.S. Senate on July 11, 1984, when my father called and said Nancy was fading. I rushed back to her bedside. My mother and father were there in the room, along with Tipper, and of course Frank

Hunger, Nancy's beloved husband.

Her pain had long since reached such agonizing levels that she was getting large doses of morphine and other painkillers, which inevitably affected her level of consciousness. Before I got back to her room, she'd been in what the other family members had described as a haze produced by the painkillers, her eyes glassy and unfocused. But the instant I walked in and Nancy heard my voice, she turned her head toward me and popped out of the fog completely, focusing intensely on my eyes. I'll never forget that moment. It was just us again: a brother and sister who could communicate without words. And I imagined—actually, I really don't think I was imagining but instead was "hearing" from her quite clearly—a powerful yet silent question: "Do you bring hope?"

I looked into her eyes and said, "I love you, Nancy." I knelt by the bed, holding her hand for a long time, and soon she took her last breath and slipped away.

I spoke publicly about my sister's death for the very first time when I accepted the Democratic nomination for vice president in 1996, and was surprised that some felt the remarks were mawkish. Nancy played too important a role in my life not to talk about her at a moment when the nation was in the midst of a struggle with the tobacco companies to get them to change their ways and stop convincing young women and men to make the same deadly mistake Nancy had made when she started smoking at 13. I wanted her story to sound an alarm about the perils of tobacco and help put a stop to the way tobacco's apologists were drowning out the sober voices of science.

But beyond that, there's no understanding me without understanding Nancy. She was a powerful force in my life when she was alive and remains so even now. Nancy was a strong, independent woman; I know that no one forced her to keep smoking all those years. But I'm also convinced that had the cigarette industry not glamorized smoking and whitewashed its ravages, my Nancy would be alive today. I wouldn't have to miss her every day of my life. I would still have her smiles and her teasing, her advice and counsel, and hugs from my loving, warm, irreplaceable big sister.

I also wish my family had extricated itself from growing tobacco sooner than it did after Nancy's illness. Truthfully, we all were numb during the onslaught of the cancer, and then our attention was focused on getting her well. The implications of continuing to grow a crop on my father's farm that helped produce the cigarettes that had caused her fatal disease seemed a little abstract and a little remote at that point—in the same way that global warming seems remote to many right now. But conversations about shutting down tobacco growing on our farm had begun when she first got sick, and not long after her death, my father decided he would stop growing tobacco altogether.

I know from this experience that it sometimes takes time to connect all the dots when accepted habits and behaviors are first found to be harmful. But I also learned that a day of reckoning might come when you very much wish that you had connected the dots more quickly.

Now, of course, just as the scientists of 1964 clearly told us that smoking kills people by causing lung cancer and other diseases, the best scientists of the 21st century are telling us ever more urgently that the global warming pollution we're pumping into Earth's atmosphere is harming the planet's climate and putting the future of human civilization at grave risk. And once again, we are taking our time—too much time—in connecting the dots.

TOP: *Al and Nancy, at the Gore Farm, Carthage, TN, 1951;* BOTTOM: *Al, Nancy, and their parents on Gore Farm, 1951*

The second problem in the way we think about the climate crisis is the wide gulf between what C. P. Snow described as "the two cultures." Science has become so specialized in its single-minded pursuit of ever-more refined knowledge in narrowing subspecialties that the rest of us have more and more difficulty making sense of their conclusions and translating their wisdom into plain language. Moreover, since science thrives on uncertainty and politics is paralyzed by it, scientists have a difficult time sounding the alarm bells for politicians, because even when their findings make it clear that we're in grave danger, their first impulse is to replicate the experiment to see if they get the same results.

Politicians, on the other hand, often confuse self-interested arguments paid for by lobbyists and planted in the popular press with legitimate, peer-reviewed studies published in reputable scientific journals. For example, the so-called global warming skeptics cite one article more than any other in arguing that global warming is just a myth: a statement of concern during the 1970s that the world might be in danger of entering a new ice

CONSENSUS AS STRO HAS DEVELOPED ARO RARE IN SCIENCE.

DONALD KENNEDY, EDITOR IN CHIEF, *SCIENCE* MAGAZINE

age. But the article in which that scientist's comment appeared was published in *Newsweek* and never appeared in any peer-reviewed journal. Moreover, the scientist who made the statement corrected it shortly thereafter with a clear explanation of why his offhand comment was erroneous.

There is a misconception that the scientific community is in a state of disagreement about whether global warming is real, whether human beings are the principal cause, and whether its consequences are so dangerous as to warrant immediate action. In fact, there is virtually no serious disagreement remaining on any of these central points that make up the consensus view of the world scientific community.

According to Jim Baker, when he was head of NOAA, the scientific agency responsible for most of the measurements related to global warming, "There is a better scientific consensus on this issue than any other...with the possible exception of Newton's Law of Dynamics." Donald Kennedy summarized this point when he said of the consensus on global warming:

NG AS THE ONE THAT UND THIS TOPIC IS

A University of California at San Diego scientist, Dr. Naomi Oreskes, published in *Science* magazine a massive study of every peer-reviewed science journal article on global warming from the previous 10 years. She and her team selected a large random sample of 928 articles representing almost 10% of the total, and carefully analyzed how many of the articles agreed or disagreed with the prevailing consensus view. About a quarter of the articles in the sample dealt with aspects of global warming that did not involve any discussion of the central elements of the consensus. Of the three-quarters that did address these main points, the percentage that disagreed with the consensus? Zero.

Number of peer-reviewed articles dealing with "climate change" published in scientific journals during the previous 10 years:

928

Percentage of articles in doubt as to the cause of global warming:

0%

Every doctor in private practice was asked:
—family physicians, surgeons, specialists...
doctors in every branch of medicine—
"What cigarette do you smoke?"

According to a recent Nationwide survey:

More Doctors
Smoke Camels

than any other cigarette!

Not a guess, not just a trend . . . but an *actual fact* based on the statements of doctors themselves to 3 nationally known independent research organizations.

The misconception that there is serious disagreement among scientists about global warming is actually an illusion that has been deliberately fostered by a relatively small but extremely well-funded cadre of special interests, including Exxon Mobil and a few other oil, coal, and utilities companies. These companies want to prevent any new policies that would interfere with their current business plans that rely on the massive unrestrained dumping of global warming pollution into the Earth's atmosphere every hour of every day.

One of the internal memos prepared by this group to guide the employees they hired to run their disinformation campaign was discovered by the Pulitzer Prize–winning reporter Ross Gelbspan. Here was the group's stated objective: to "reposition global warming as theory, rather than fact."

This technique has been used before.

The tobacco industry, 40 years ago, reacted to the historic Surgeon General's report linking cigarette smoking to lung cancer and other lung diseases by organizing a similar disinformation campaign. One of their memos, prepared in the 1960s, was recently uncovered during one of the lawsuits against the tobacco companies on behalf of the millions of people who have been killed by their product. It is interesting to read it 40 years later in the context of the ongoing global warming disinformation campaign:

"Doubt is our product, since it is the best means of competing with the 'body of fact' that exists in the mind of the general public. It is also the means of establishing a controversy." Brown and Williamson Tobacco Company memo, 1960s

One prominent source of disinformation on global warming has been the Bush-Cheney White House.

They have attempted to silence scientists working for the government who, like James Hansen at NASA, have tried to warn about the extreme danger we are facing. They have appointed "skeptics" recommended by oil companies to key positions, from which they can prevent action against global warming. As our principal negotiators in international forums, these skeptics can prevent agreement on a worldwide response to global warming.

At the beginning of 2001, President Bush hired a lawyer/lobbyist named Phillip Cooney to be in charge of environmental policy in the White House. For the previous six years, Cooney had worked at the American Petroleum Institute and was the person principally in charge of the oil and coal companies' campaign to confuse the American people about this issue.

Even though Cooney has no scientific training whatsoever, he was empowered by the president to edit and censor the official assessments of global warming from the EPA and other parts of the federal government. In 2005, a White House memo authorized by Cooney (a portion of which appears below) was leaked to the *New York Times* by a hidden whistleblower inside the administration. Cooney had diligently edited out any mention of the dangers global warming poses to the American people. The newspaper's disclosure was embarrassing to the White House, and Cooney, in what has become a rare occurrence in the last few years, resigned. The next day, he went to work for Exxon Mobil.

The New York Times

~~Warming will also cause reductions in mountain glaciers and advance the timing of the melt of mountain snow peaks in polar regions. In turn, runoff rates will change and flood potential will be altered in ways that are currently not well understood. There will be significant shifts in the seasonality of runoff that will have serious impacts on native populations that rely on fishing and hunting for their livelihood. These changes will be further complicated by shifts in precipitation regimes and a possible intensification and increased frequency of hydrologic events.~~ Reducing the uncertainties in current understanding of the relationships between climate change and Arctic hydrology is critical.

straying from research strategy into speculative findings from here.

Has the disinformation campaign on global warming succeeded?

Well, alongside the study of peer-reviewed scientific journal articles that showed 0% in disagreement with the consensus on global warming, another large study was conducted of all the articles on global warming during the previous 14 years in the four newspapers considered by the authors of the study to be the most influential in America: the *New York Times*, the *Washington Post*, the *LA Times*, and the *Wall Street Journal*.

They selected a large random sample of almost 18% of the articles. Astonishingly, they found that more than one-half gave equal weight to the consensus view on the one hand, and the scientifically discredited view that human beings play no role in global warming on the other. The authors concluded that American news media had been falsely "giving the impression that the scientific community was embroiled in a rip-roaring debate on whether or not humans were contributing to global warming."

No wonder people are confused.

Articles in the popular press about global warming during the previous 14 years:

636

Percentage of articles in doubt as to the cause of global warming:

53%

Phillip Cooney

1995–JANUARY 20, 2001

American Petroleum Institute lobbyist in charge of global warming disinformation

JANUARY 20, 2001

Hired as chief of staff, White House Environment Office

JUNE 14, 2005

Leaves White House to go on payroll of Exxon Mobil

A century ago, Upton Sinclair was one of the most respected members of a remarkable group of investigative journalists and authors who uncovered horrendous abuses that had been hidden in the excesses of the Gilded Age and helped to stimulate the reforms of the Progressive Era. Sinclair made a point then that might well be applied to the group of naysayers appointed by the Bush-Cheney administration to be in charge of America's response to global warming—naysayers like Cooney who are working to convince Americans that the problem is not real, much less dangerous, and that we're not responsible for it in any way.

IT IS DIFFICULT TO GET UNDERSTAND SOMET SALARY DEPENDS UP UNDERSTANDING IT.

UPTON SINCLAIR

A MAN TO
HING WHEN HIS
ON HIS NOT

What makes this kind of dishonesty intolerable is that there is so much at stake.

On June 21, 2004, 48 Nobel Prize—winning scientists accused President Bush and his administration of distorting science:

'By ignoring scientific consensus on critical issues such as global climate change, [President Bush and his administration] are threatening the Earth's future."

SIGNED BY:

Peter Agre
CHEMISTRY 2003

Sidney Altman
CHEMISTRY 1989

Philip W. Anderson
PHYSICS 1977

David Baltimore
MEDICINE 1975

Baruj Benacerraf
MEDICINE 1980

Paul Berg
CHEMISTRY 1980

Hans A. Bethe
PHYSICS 1967

Michael Bishop
MEDICINE 1989

Günter Blobel
MEDICINE 1999

N. Bloembergen
PHYSICS 1981

James W. Cronin
PHYSICS 1980

Johann Deisenhofer
CHEMISTRY 1988

John B. Fenn
CHEMISTRY 2002

Val Fitch
PHYSICS 1980

Jerome I. Friedman
PHYSICS 1990

Walter Gilbert
CHEMISTRY 1980

Alfred G. Gilman
MEDICINE 1994

Donald A. Glaser
PHYSICS 1960

Sheldon L. Glashow
PHYSICS 1979

Joseph Goldstein
MEDICINE 1985

Roger Guillemin
MEDICINE 1977

Dudley Herschbach
CHEMISTRY 1986

Roald Hoffmann
CHEMISTRY 1981

H. Robert Horvitz
MEDICINE 2002

David H. Hubel
MEDICINE 1981

Louis Ignarro
MEDICINE 1998

Eric R. Kandel
MEDICINE 2000

Walter Kohn
CHEMISTRY 1998

Arthur Kornberg
MEDICINE 1959

Leon M. Lederman
PHYSICS 1988

Tsung-Dao Lee
PHYSICS 1957

David M. Lee
PHYSICS 1996

William N. Lipscomb
CHEMISTRY 1976

Roderick MacKinnon
CHEMISTRY 2003

Mario J. Molina
CHEMISTRY 1995

Joseph E. Murray
MEDICINE 1990

Douglas D. Osheroff
PHYSICS 1996

George Palade
MEDICINE 1974

Arno Penzias
PHYSICS 1978

Martin L. Perl
PHYSICS 1995

Norman F. Ramsey
PHYSICS 1989

Burton Richter
PHYSICS 1976

Joseph H. Taylor Jr.
PHYSICS 1993

E. Donnall Thomas
MEDICINE 1990

Charles H. Townes
PHYSICS 1964

Harold Varmus
MEDICINE 1989

Eric Wieschaus
MEDICINE 1995

Robert W. Wilson
PHYSICS 1978

The third problem in our way of thinking about global warming is our false belief that we have to choose between a healthy economy and a healthy environment.

In 1991 I was part of a bipartisan group of senators who tried to convince the first Bush administration to go to the Earth Summit in Rio de Janeiro. In response, the White House organized a conference to communicate the impression that they were behaving responsibly. And, as part of their effort, they made a colorful pamphlet about "Global Stewardship" to persuade people that they were acting to protect the global environment.

I was particularly intrigued by one of the images depicting the way they saw the balance between the environment and the economy.

The illustration is emblematic of a widely held view concerning the fundamental "choice" that the United States must make between the economy and the environment. The image features an old-fashioned set of scales. On one side are gold bars, representing wealth and economic success. On the other side is...the entire planet!

USING MARKET CAPITALISM AS AN ALLY

One of the keys to solving the climate crisis involves finding ways to use the powerful force of market capitalism as an ally. And more than anything else, that requires accurate measurements of the real consequences—positive and negative—of all the important economic choices we make.

The environmental impact of our economic choices has often been ignored because traditional business accounting has allowed these factors to be labeled "externalities" and routinely excluded from the balance sheet. It is not suprising that this unwise practice has persisted for as long as it has. These factors are sometimes difficult to put a price tag on. And, by simply declaring that these factors are "external," it is easy to put them out of sight and out of mind.

Now, however, many business leaders are finally recognizing the full effects of their choices and "price in" such factors as the environment, community impact, and employee longevity by using sophisticated techniques to measure their true value.

Part of this strategy involves taking a broader view of how businesses can sustain their profitability over time. These leaders are abandoning their obsessive short-term focus for a longer-term view. This can often make a big difference in evaluating the pros and cons of investments that are likely to pay off after two or three years. Many such investments are routinely avoided today because the marketplace penalizes any expenditures that hurt short-term profits.

But there is also a big change under-way in the investment community, led by investors who have become dissatisfied with the "short-termism" in the financial markets and who want to adopt a more realistic view of how businesses build up and retain their value. These investors are taking the environment and other factors into account when they evaluate an investment. For example, many individual and institutional investors are now deciding that it is prudent to consider the potential impact of their investments on climate change.

Whether it is putting money into a savings account at a bank or local credit union, buying stocks, investing in mutual funds for your retirement, or managing your kid's college fund, it matters where your money goes. Considering sustainability issues when investing need not diminish returns—indeed, some evidence shows that it can enhance them. You can make a contribution to stopping climate change, support global sustainability, and do well financially if you choose your investments wisely.

The implication is that this is not only a choice we have to make, but a difficult one. But, in fact, it's a false choice for two reasons. First, without a planet, we won't fully enjoy those gold bars.

And second, if we do the right thing, then we're going to create a lot of wealth, jobs, and opportunity.

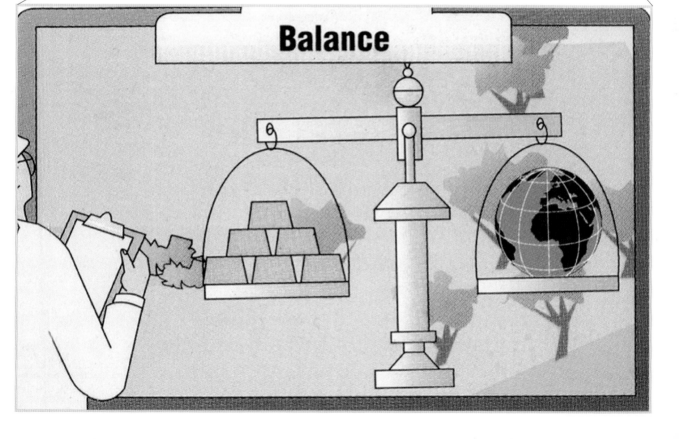

Unfortunately, the false choice posed between our economy and the environment affects our policies in harmful ways.

One example is auto mileage standards. Japan has cars that are required by law to get more than 45 miles per gallon. Europe is not far behind, and has passed new laws designed to surpass Japanese standards. Our friends in Canada and Australia are moving toward higher requirements of more than 30 miles per gallon.

Yet the United States is dead last.

We're told that we have to protect our automobile companies from competition in places like China where, it is said, their leaders don't care about the environment.

In fact, Chinese emissions standards have been raised and already far exceed our own. Ironically, we cannot sell cars made in America to China because we don't meet their environmental standards.

COMPARISON OF FUEL ECONOMY AND GHG EMISSIONS STANDARDS AROUND THE WORLD

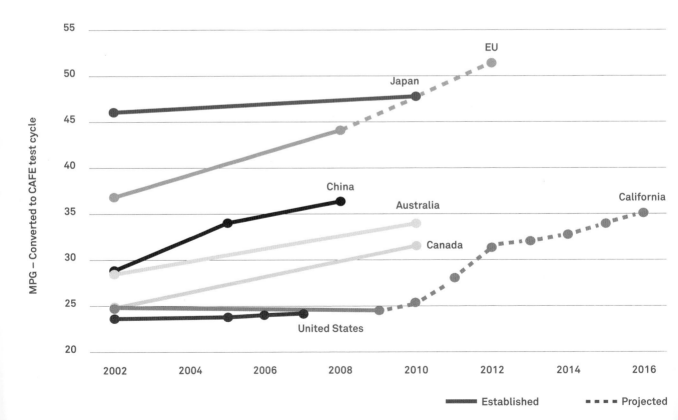

In California, the state legislature has taken the initiative to require higher standards for cars sold in California. But the auto companies are suing California to prevent this state law from taking effect—because it would mean that, *10 years from now*, they would have to manufacture cars for California that are almost as efficient as China is making today.

Our outdated environmental standards are based on faulty thinking about the true relationship between the economy and the environment. They are intended in this case to help automobile companies succeed. But as the chart makes clear, it's the companies building more efficient cars that are doing well. The U.S. companies are in deep trouble. And they're still redoubling their efforts to sell large, inefficient gas-guzzlers even though the marketplace is sending the same message that the environment is sending.

CHANGE IN MARKET CAPITALIZATION: FEBRUARY–NOVEMBER 2005

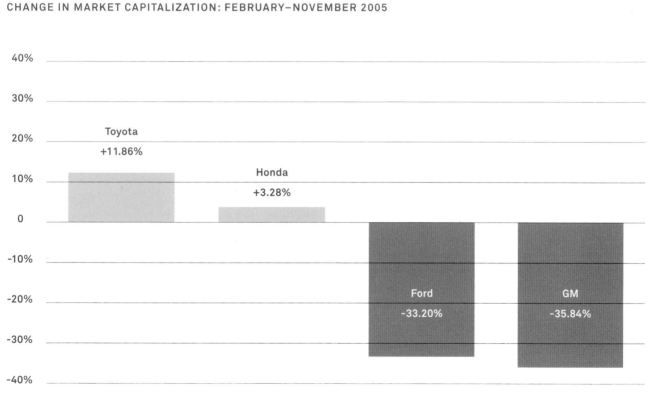

SOURCE: FORBES.COM

Luckily, more and more U.S. business executives are beginning to lead us in the right direction.

For example, General Electric recently announced a dramatic new initiative on global warming. Jeffrey Immelt, the CEO of GE, explained how the environment and the economy fit together in his vision:

WE THINK GREEN ME A TIME PERIOD WHER IMPROVEMENT IS GOI PROFITABILITY.

JEFFREY R. IMMELT, CHAIRMAN AND CEO, GE

ANS GREEN. THIS IS
E ENVIRONMENTAL
NG TO LEAD TOWARD

The fourth and final problem in the way some people think about global warming is the dangerous misconception that if it really is as big a threat as the scientists are telling us it is, then maybe we're helpless to do anything about it so we might as well throw up our hands.

An astonishing number of people go straight from denial to despair, without pausing on the intermediate step of saying, "We can do something about this!"

U.S. RENEWABLE ENERGY FUTURE

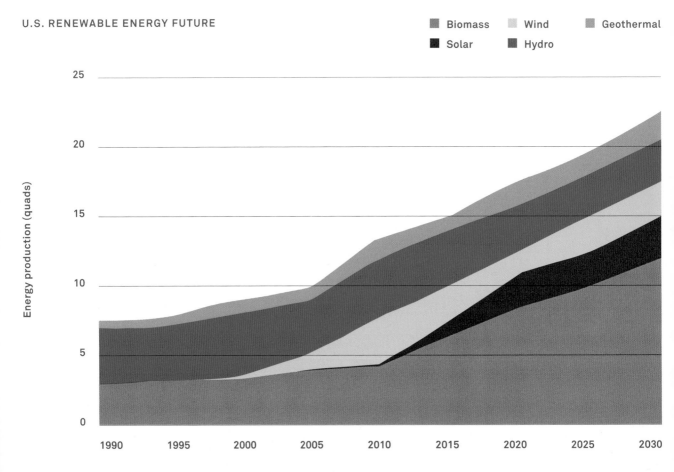

And we can.

COMPACT FLUORESCENT LIGHTBULBS

FUEL-CELL HYBRID BUSES

SOLAR PANELS

GREEN ROOF

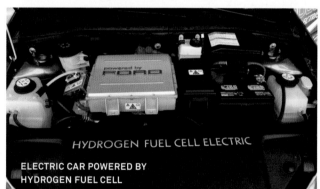
HYDROGEN FUEL CELL ELECTRIC

ELECTRIC CAR POWERED BY
HYDROGEN FUEL CELL

HYBRID CAR

GEOTHERMAL POWER STATION

We have everything we need to begin solving the climate crisis—save, perhaps, political will. But in America, political will is a renewable resource.

Each one of us is a cause of global warming, but each of us can become part of the solution: in the decisions we make on what we buy, the amount of electricity we use, the cars we drive, and how we live our lives. We can even make choices to bring our individual carbon emissions to zero.

MIDDELGRUNDEN OFFSHORE WIND FARM, COPENHAGEN, DENMARK, 2001

WIND POWER

Without wind power we could never have settled the Great Plains. For all the credit we give to railroads, rifles, and horses, it was the windmill that for generations tirelessly pumped underground water to the surface, helping settlers cook, wash, and tend their livestock.

Wind has always been a resource waiting to be tapped. A 100-megawatt wind farm—that's 50 300-foot towers carrying two-megawatt turbines the size of a tractor trailer truck—can power 24,000 homes. You would have to burn nearly 50,000 tons of coal to provide the same amount of electricity. You have to burn the same amount of coal each year to generate the same amount of energy: imagine the loads of carbon dioxide that this produces annually.

It's true that a modern wind turbine also releases carbon, but only during its manufacture; once it's up and running the turbine runs clean. The comparison between coal and wind as energy sources is stark: While coal spews a constant stream of Earth-warming carbon, wind power emits none.

The market has already decided that wind generation is one of the most mature and cost-effective technologies available to power our future. Utilities all over the country are investing in wind farms. In 2005 General Electric's turbine business doubled. Vestas, the global leader in this area, has made windmills into Denmark's largest export. On some nights along the Danish coast, the winter winds meet all of the local energy needs. By 2008 a quarter of that country's electricity will be pulled from the sky.

It's true these windmills are huge, but then so is our appetite for electricity. They alter our skylines, but many find watching their spinning blades peaceful to look at.

Every day we continue to fill the air with carbon exhaust while wind power is right there, waiting to be tapped.

Two economists at Princeton University—Robert Socolow and his colleague Stephen Pacala—concluded in a respected study of policies that can help us solve the climate crisis: "Humanity already possesses the fundamental scientific, technical, and industrial know-how to solve the carbon and climate problems for the next half-century."

The graphic below, which has been based on the Socolow/Pacala study, illustrates at the top right how much U.S. global warming pollution is expected to increase over the next four decades if we continue "business as usual." But each of the colored wedges shows the reduction in pollution that can be accomplished in the same period by using the six sets of new policies described here.

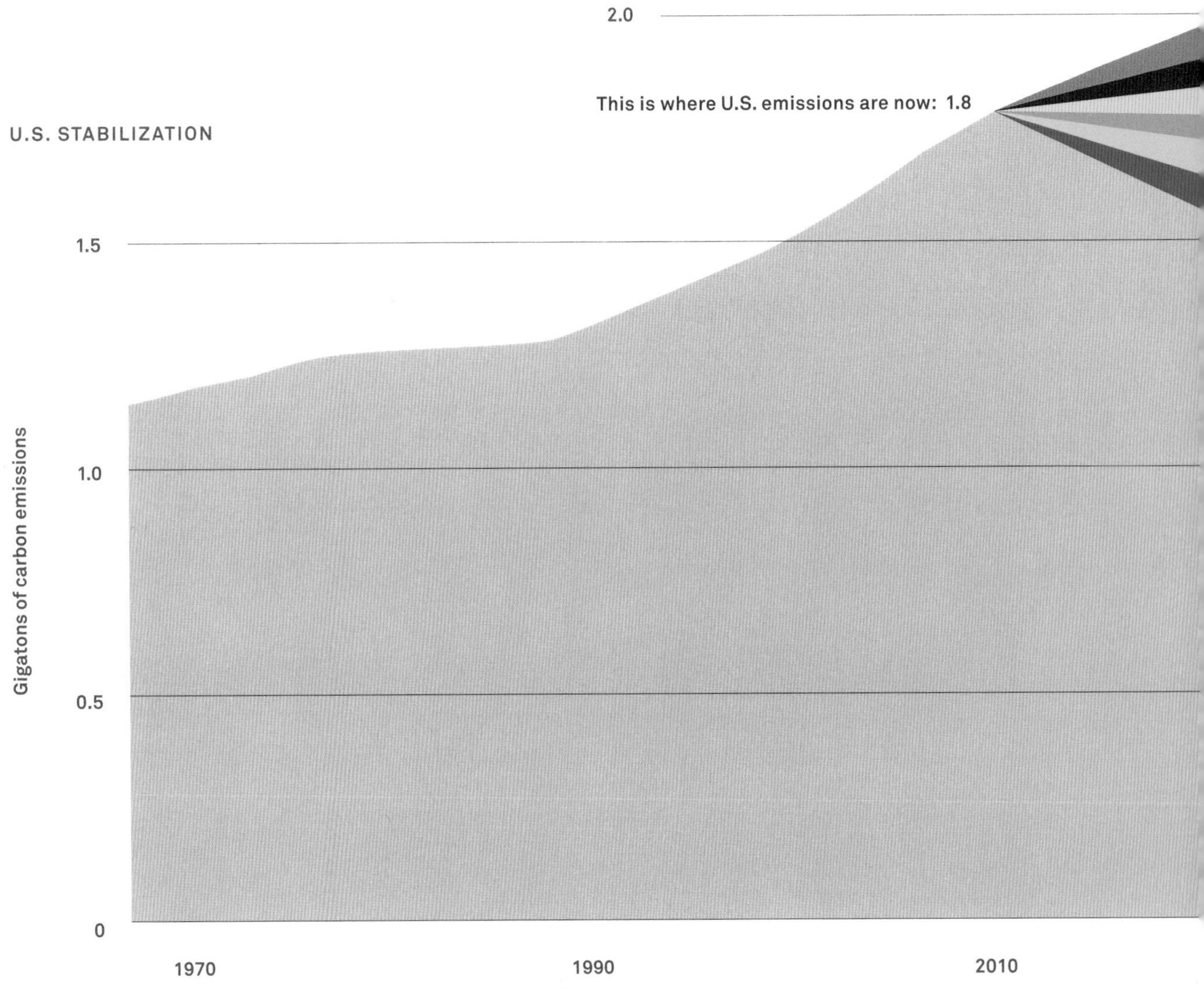

U.S. STABILIZATION

2.0

This is where U.S. emissions are now: 1.8

1.5

Gigatons of carbon emissions

1.0

0.5

0

1970   1990   2010

Together, these changes, all of which are based on already-existing, affordable technologies, can bring emissions down to a point below 1970s levels.

Business as usual takes them way up here: 2.6

2.5

Reduction from more efficient use of electricity in heating and cooling systems, lighting, appliances, and electronic equipment

Reduction from end-use efficiency, meaning that we design buildings and businesses to use far less energy than they currently do

Reduction from increased vehicle efficiency by manufacturing cars that run on less gas and putting more hybrid and fuel-cell cars on the roads

Reduction from making other changes in transport efficiency, such as designing cities and towns to have better mass transit systems and building heavy trucks that have greater fuel efficiency

Reduction from increased reliance on renewable energy technologies that already exist, such as wind and biofuels

Reduction from the capture and storage of excess carbon from power plants and industrial activities

2030   2050

Other countries have already decided to act. The Kyoto Treaty has now been ratified by 132 nations in the developed world.

There are only two advanced nations that have not ratified Kyoto, and we are one of them. The other is Australia.

RATIFIED BY

Algeria
Antigua
Arabia
Argentina
Armenia
Austria
Azerbaijan
Bahamas
Bangladesh
Barbados
Barbuda
Belgium
Belize
Benin
Bhutan
Bolivia
Botswana
Brazil
Bulgaria
Burundi
Cambodia
Cameroon
Canada
Chile
China
Colombia

Cook Islands
Costa Rica
Cuba
Cyprus
Czech Republic
Denmark
Djibouti
Dominica
Dominican Republic
Ecuador
Egypt
El Salvador
Equatorial Guinea
Estonia
European Union
Fiji
Finland
France
Gambia
Georgia
Germany
Ghana
Greece
Grenada
Guatemala
Guinea
Guyana

Honduras
Hungary
Iceland
India
Indonesia
Ireland
Israel
Italy
Jamaica
Japan
Jordan
Kenya
Kiribati
Kyrgyzstan
Laos
Latvia
Lesotho
Liberia
Liechtenstein
Lithuania
Luxembourg
Macedonia
Madagascar
Malawi
Malaysia
Maldives
Mali

Are we going to be left behind as the rest of the world moves forward?

Malta
Marshall Islands
Mauritius
Mexico
Micronesia
Mongolia
Morocco
Mozambique
Myanmar
Namibia
Nauru
Netherlands
New Zealand
Nicaragua
Niger
Nigeria
Niue
North Korea
Norway
Oman
Pakistan
Palau
Panama
Papua New Guinea
Paraguay
Peru
Philippines

Poland
Portugal
Qatar
Republic of Moldova
Romania
Russian Federation
Rwanda
Saint Lucia
Samoa
Saudi Arabia
Senegal
Seychelles
Slovakia
Slovenia
Solomon Islands
South Africa
South Korea
Spain
Sri Lanka
St. Vincent & Grenadines
Sudan
Sweden
Switzerland
Tanzania
Thailand
Togo
Trinidad & Tobago

Tunisia
Turkmenistan
Tuvalu
United Arab Emirates
Uganda
Ukraine
United Kingdom
Uruguay
Uzbekistan
Vanuatu
Venezuela
Vietnam
Yemen

NOT RATIFIED BY
Australia
United States

The Politicization of Global Warming

As I've traveled around the world giving my slide show, there are two questions I most often get—particularly in the United States—from people who already know how serious the crisis has become are:

(1) "Why do so many people still believe this crisis isn't real?" and

(2) "Why is this a political issue at all?"

My response to the first question has been to try to make my slide show—and now this book—as clear and compelling as I can. As for why so many people still resist what the facts clearly show, I think, in part, the reason is that the truth about the climate crisis is an inconvenient one that means we are going to have to change the way we live our lives. Most of these changes will turn out to be for the better—things we really should do for other reasons anyway—but they are inconvenient nonetheless. Whether these changes involve something as minor as adjusting the thermostat and using different light bulbs, or as major as switch-

ing from oil and coal to renewable fuels, they will require effort.

But the answer to the first question is also linked to the second question. The truth about global warming is especially inconvenient and unwelcome to some powerful people and companies making enormous sums of money from activities they know full well will have to change dramatically in order to ensure the planet's livability.

These people—especially those at a few multinational companies with the most at stake—have been spending many millions of dollars every year in figuring out ways of sowing public confusion about global warming. They've been particularly effective in building a coalition with other groups who agree to support each other's interests, and that coalition has thus far managed to paralyze America's ability to respond to global warming. The Bush/Cheney administration has received strong support from

Al Gore in the Old Executive Office Building, Washington, DC, 1999

this coalition and seems to be doing everything it can to satisfy their concerns.

For example, many scientists working on global-warming research throughout the government have been ordered to watch what they say about the climate crisis and instructed not to talk to the news media. More important, all of America's policies related to global warming have been changed to reflect the unscientific view—the administration's view—that global warming is not a problem. Our negotiators in international forums dealing with global warming have been advised to try and stop any movement toward action that would inconvenience oil or coal companies, even if this means disrupting the diplomatic machinery in order to do it.

own party becoming less concerned, probably because they're naturally more inclined to give the president the benefit of the doubt.

The rationale offered by the so-called global warming skeptics for opposing any action to solve the climate crisis has changed several times over the years. At first, opponents argued that global warming was not occurring at all; they said it was just a myth. A few of them still say that today, but now there is so much undeniable evidence demolishing that assertion that, most naysayers have decided they need to change tactics. They now acknowledge that the globe is indeed warming, but in the very next breath, they claim it is just due to "natural causes."

edge that the crisis it's causing is real and harmful. Their philosophy seems to be "eat, drink, and be merry, for tomorrow our children will inherit the worst of this crisis; it's too inconvenient for us to be bothered."

All of these shifting rationales usually rely on the same underlying political tactic: Assert that the science is uncertain and that there is grave doubt about the underlying facts.

These groups emphasize uncertainty because they know that politics in America can be paralyzed by it. They understand that it is a politician's natural instinct to avoid taking any stand that seems controversial unless and until the voters demand it or conscience absolutely requires it. So if voters and the politicians

The truth about the climate crisis is an inconvenient one that means we are going to have to change the way we live our lives.

In addition, President Bush appointed the person in charge of the oil company disinformation campaign on global warming to head up all environmental policy in the White House. Even though this lawyer/lobbyist had no scientific training whatsoever, he was empowered by the president to edit and censor all warnings from the EPA and other government agencies about global warming.

Political leaders—especially the president—can have a major effect not only on public policy (especially when Congress is controlled by the president's party, is compliant, and does whatever the president wants it to) but also on public opinion, especially among those who count themselves followers of the president.

Consider this fact: Even as Americans in general have become increasingly concerned about global warming, opinion polls show members of the president's

President Bush himself still tries to take that position, asserting that even though it does seem the world is getting warmer, hey, there's no compelling evidence that human beings are responsible for it. And he seems to be particularly certain that the oil and coal companies that so strongly support him couldn't possibly have anything to do with it.

Another related argument used by the deniers is that yes, global warming does seem to be happening, but it will probably be good for us. Certainly any effort to stop it, they continue, would no doubt be bad for the economy.

But the latest—and in my opinion, most disgraceful—argument put forth by opponents of change is: Yes, it's happening, but there's nothing we can really do about it, so we might as well not even try. This faction favors the continued dumping of global-warming pollution into the atmosphere, even though they acknowl-

who represent them can be convinced that scientists themselves disagree on fundamental issues concerning global warming, then the political process can be paralyzed indefinitely. That is exactly what has happened—at least until quite recently—and it is still unclear when the situation will really change.

Part of the problem has to do with a long-term structural change in the way America's marketplace of ideas now operates. The one-way nature of our dominant communications medium, television, has combined with the increasing concentration of ownership over the vast majority of media outlets by a smaller and smaller number of large conglomerates that mix entertainment values with journalism to seriously damage the role of objectivity in America's public forum. Today there are many fewer independent journalists with the freedom and stature to blow the whistle when important

Gore speaking on Earth Day, 1997

facts are consistently being distorted in order to deceive the public. The Internet offers the most hopeful opportunity to restore integrity to the public dialogue, but television is still dominant in shaping that dialogue.

The "propaganda" techniques that emerged with the new film and broadcasting mass media of the 20th century prefigured the widespread use of related techniques for mass advertising and for political persuasion. And now, corporate lobbying efforts to influence and control public policy have been stepped up dramatically, which in turn is leading to the widespread and often cynical use of these same mass persuasion techniques to condition the public's thinking about important issues lest they begin to support solutions that will be inconvenient—and expensive—for particular industries.

One of the persistent techniques in the campaign to stop action against the climate crisis has been to repeatedly and persistently accuse the scientists trying to warn us about the crisis of being dishonest, greedy, and untrustworthy and of misrepresenting scientific facts in order to somehow beef up their research grants.

These charges are insulting and ludicrous, but they have been repeated often enough and loudly enough—through so many media megaphones—that many people do now wonder if the charges are true. And that is particularly ironic, given that so many of the skeptics actually do receive funding and support from self-interested groups financed by corporations desperate to stop any action against global warming. Incredibly, the public has been hearing the discredited views of these skeptics as much as or more than they have heard the consensus view of the global scientific community. That disgraceful fact is a notable stain on the record of America's modern news media, and many leaders of journalism are belatedly taking steps to correct it.

But it is far from clear that the news media will be able to sustain a higher commitment to objectivity in the face of the intense pressures that increasingly erode it and render it shockingly vulnerable to this kind of organized propaganda. We have lost a lot of time that could have been spent solving the crisis, because the opponents of action have thus far successfully politicized the issue in the minds of many Americans.

We can't afford inaction any longer, and, frankly, there's just no excuse for it. We all want the same thing: for our children and the generations after them to inherit a clean and beautiful planet capable of supporting a healthy human civilization. That goal should transcend politics.

Yes, the science is ongoing and always evolving, but there's already enough data, enough damage, to know without question that we're in trouble. This isn't an ideological debate with two sides, pro and con. There is only one Earth, and all of us who live on it share a common future. Right now we are facing a planetary emergency, and it is time for action, not for more phony controversies designed to insure political paralysis.

Many U.S. cities have "ratified" the Kyoto Treaty on their own and are implementing policies to reduce global warming pollution below the levels required by the protocol.

ARKANSAS
Fayetteville
Little Rock
North Little Rock

CALIFORNIA
Albany
Aliso Viejo
Arcata
Berkeley
Burbank
Capitola
Chino
Cloverdale
Cotati
Del Mar
Dublin
Fremont
Hayward
Healdsburg
Hemet
Irvine
Lakewood
Los Angeles
Long Beach
Monterey Park
Morgan Hill
Novato
Oakland
Palo Alto
Petaluma
Pleasanton
Richmond
Rohnert Park

Sacramento
San Bruno
San Francisco
San Luis Obispo
San Jose
San Leandro
San Mateo
Santa Barbara
Santa Cruz
Santa Monica
Santa Rosa
Sebastopol
Sonoma
Stockton
Sunnyvale
Thousand Oaks
Vallejo
West Hollywood
Windsor

COLORADO
Aspen
Boulder
Denver
Telluride

CONNECTICUT
Bridgeport
Easton
Fairfield
Hamden
Hartford
Mansfield
Middletown
New Haven

Stamford

DELAWARE
Wilmington

FLORIDA
Gainesville
Hallandale Beach
Holly Hill
Hollywood
Key Biscayne
Key West
Lauderhill
Miami
Miramar
Pembroke Pines
Pompano Beach
Port St. Lucie
Sunrise
Tallahassee
Tamarac
West Palm Beach

GEORGIA
Atlanta
Athens
East Point
Macon

HAWAII
Hilo
Honolulu
Kauai
Maui

ILLINOIS
Carol Stream
Chicago

Highland Park
Schaumburg
Waukegan

INDIANA
Columbus
Fort Wayne
Gary
Michigan City

IOWA
Des Moines

KANSAS
Lawrence
Topeka

KENTUCKY
Lexington
Louisville

LOUISIANA
Alexandria
New Orleans

MARYLAND
Annapolis
Baltimore
Chevy Chase

MASSACHUSETTS
Boston
Cambridge
Malden
Medford
Newton
Somerville
Worcester

MICHIGAN
Ann Arbor

Grand Rapids

Southfield

MINNESOTA

Apple Valley

Duluth

Eden Prairie

Minneapolis

St. Paul

MISSOURI

Clayton

Florissant

Kansas City

Maplewood

St. Louis

Sunset Hills

University City

MONTANA

Billings

Missoula

NEBRASKA

Bellevue

Lincoln

Omaha

NEVADA

Las Vegas

NEW HAMPSHIRE

Keene

Manchester

Nashua

NEW JERSEY

Bayonne

Bloomfield

Brick Township

Elizabeth

Hamilton

Hightstown

Hope

Hopewell

Kearny

Newark

Plainfield

Robbinsville

Westfield

NEW MEXICO

Albuquerque

NEW YORK

Albany

Buffalo

Hempstead

Ithaca

Mt. Vernon

New York City

Niagara Falls

Rochester

Rockville Centre

Schenectady

White Plains

NORTH CAROLINA

Asheville

Chapel Hill

Durham

OHIO

Brooklyn

Dayton

Garfield Heights

Middletown

Toledo

OKLAHOMA

Norman North

OREGON

Corvallis

Eugene

Lake Oswego

Portland

PENNSYLVANIA

Erie

Philadephia

RHODE ISLAND

Pawtucket

Providence

Warwick

SOUTH CAROLINA

Charleston

Sumter

TEXAS

Arlington

Austin

Denton

Euless

Hurst

Laredo

McKinney

UTAH

Moab

Park City

Salt Lake City

VERMONT

Burlington

VIRGINIA

Alexandria

Charlotteville

Virginia Beach

WASHINGTON

Auburn

Bainbridge Island

Bellingham

Burien

Edmonds

Issaquah

Kirkland

Lacey

Lynnwood

Olympia

Redmond

Renton

Seattle

Tacoma

Vancouver

WASHINGTON, DC

WISCONSIN

Ashland

Greenfield

La Crosse

Madison

Racine

Washburn

Wauwatosa

West Allis

But what about the rest of us?

Ultimately, the question comes down to this: Are we, as Americans, capable of doing great things, even though they might be difficult?

Are we capable of transcending our own limitations and rising to take responsibility for charting our own destiny?

Well, the record indicates that we have this capacity.

We fought a revolution and brought forth a new nation, based on liberty and individual dignity.

We won two wars against fascism simultaneously, in the Atlantic and the Pacific, and then we won the peace that followed.

We made a moral decision that slavery was wrong, and that we could not be half-free and half-slave.

We recognized that women should have the right to vote.

We cured fearsome diseases like polio and smallpox, improved the quality of life, and extended our lifetimes.

We took on the moral challenge of desegregation and passed civil rights laws to remedy injustice against minorities.

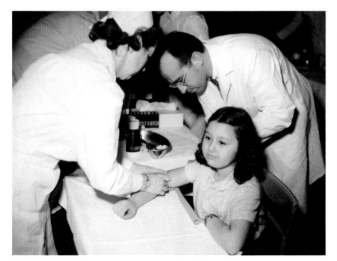

We landed on the Moon—one of the most inspiring examples of what we can do when we put our minds to it.

APOLLO 11 ASTRONAUT BUZZ ALDRIN ON THE MOON, 1969

We have even solved a global environmental crisis before. The problem of the hole in the stratospheric ozone layer was said to be impossible to fix because the causes were global and the solution required cooperation from every nation in the world. But the United States took the lead—on a bipartisan basis—with a Republican president and a Democratic Congress.

We drafted a treaty, secured worldwide agreement on it, and began to eliminate the chemicals that were causing the problem.

THE CFC SUCCESS STORY

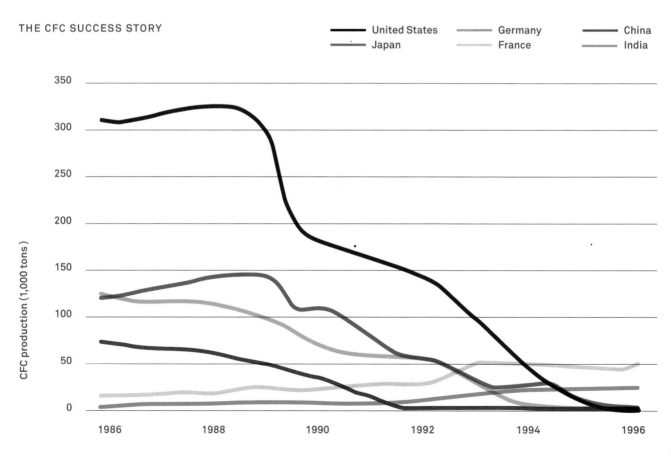

Production of chlorofluorocarbons in selected countries, 1986–1996

SOURCE: UNEP, 1999

Now, all over the world we are well on our way to solving the stratospheric ozone crisis.

HEALING THE OZONE LAYER

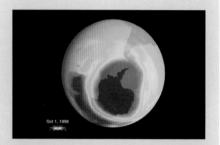

Oct 1, 1998

Once upon a time, your refrigerator could kill you. Early models used toxic and explosive gases to keep food cold. But then, in 1927, the chemist Thomas Midgley invented chlorofluorocarbons—or CFCs—to replace those gases. Touted as an innovation, CFCs revolutionized refrigeration, and eventually this seemingly harmless family of chemicals found its way into all kinds of products. As it turned out, consumers should have been suspicious. Midgley's previous claim to fame was the creation of leaded gasoline.

By 1974 millions of refrigerators containing CFCs had been sold around the world. Then two scientists began to look more closely at their impact. Dr. F. Sherwood Rowland and Dr. Mario Molina theorized that as these chemicals rose into the upper atmosphere, their molecules would be broken down by the Sun, releasing chlorine into the ozone layer and setting in motion a dangerous chain reaction.

Ozone is a simple combination of three oxygen molecules that, in the Earth's stratosphere, protects us from the Sun's most damaging rays. Rowland and Molina believed that chlorine would mix with ozone on the surface of ice particles in the stratosphere and, when hit by sunlight, would eat away at this fragile protective skin, allowing the sun's ultraviolet rays to stream unimpeded through the atmosphere, thereby damaging plant and animal health, causing skin cancer, and even threatening damage to our eyesight.

These scientists, along with Paul Crutzen, shared the Nobel Prize in 1995 for their work in atmospheric chemistry. And, more important, they sounded the alarm about ozone depletion. At first only a few environmentalists and atmospheric chemists had paid attention, but Rowland and Molina continued to make new discoveries and to refine their predictions. In 1984 a dramatic hole in the ozone layer was discovered above Antarctica, just as the scientists forecasted.

That got things moving. In 1987, 27 nations signed the Montreal Protocol, the first global environmental agreement to regulate CFCs. And as the science has improved, more and more nations have signed on. At last count there were 183. And each time they meet they strengthen its language and requirements. United Nations Secretary General Kofi Annan has called the Montreal Protocol "perhaps the single most successful international agreement to date." The impact is concrete: Since 1987 the levels of the most critical CFCs and related compounds have stabilized or declined. And while the ozone layer's full recovery will take longer than we initially thought, our efforts to date represent a significant step forward.

Controlling greenhouse gases will be more difficult, because carbon dioxide—the major greenhouse gas—is more closely linked to the global economy than CFCs ever were. Weaning our industries—and changing our personal habits—will be a challenge. But our experience with the ozone layer shows us that the people of the world actually can work together to repair some of our own mistakes, despite our often conflicting political and economic interests. Today, as the CO_2 crisis unites us, we must remember the lesson of the CFC battle: that cool heads can prevail and alter the course of environmental change for the better.

Now it is up to us to use our democracy
and our God-given ability to reason
with one another about our future and
make moral choices to change the
policies and behaviors that would, if
continued, leave a degraded, diminished,
and hostile planet for our children and
grandchildren—and for humankind.

We must choose instead to make the
21st century a time of renewal. By seizing
the opportunity that is bound up in this
crisis, we can unleash the creativity,
innovation, and inspiration that are just
as much a part of our human birthright
as our vulnerability to greed and pettiness. The choice is ours. The responsibility
is ours. The future is ours.

VIEW OF STAR-FORMING REGION
S106 IRS4 AS SEEN FROM THE
SUBARU TELESCOPE, MAUNA
KEA, HI, 2001

Now, all over the world we are well on our way to solving the stratospheric ozone crisis.

HEALING THE OZONE LAYER

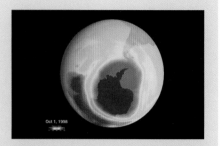

Oct 1, 1998

Once upon a time, your refrigerator could kill you. Early models used toxic and explosive gases to keep food cold. But then, in 1927, the chemist Thomas Midgley invented chlorofluorocarbons—or CFCs—to replace those gases. Touted as an innovation, CFCs revolutionized refrigeration, and eventually this seemingly harmless family of chemicals found its way into all kinds of products. As it turned out, consumers should have been suspicious. Midgley's previous claim to fame was the creation of leaded gasoline.

By 1974 millions of refrigerators containing CFCs had been sold around the world. Then two scientists began to look more closely at their impact. Dr. F. Sherwood Rowland and Dr. Mario Molina theorized that as these chemicals rose into the upper atmosphere, their molecules would be broken down by the Sun, releasing chlorine into the ozone layer and setting in motion a dangerous chain reaction.

Ozone is a simple combination of three oxygen molecules that, in the Earth's stratosphere, protects us from the Sun's most damaging rays. Rowland and Molina believed that chlorine would mix with ozone on the surface of ice particles in the stratosphere and, when hit by sunlight, would eat away at this fragile protective skin, allowing the sun's ultraviolet rays to stream unimpeded through the atmosphere, thereby damaging plant and animal health, causing skin cancer, and even threatening damage to our eyesight.

These scientists, along with Paul Crutzen, shared the Nobel Prize in 1995 for their work in atmospheric chemistry. And, more important, they sounded the alarm about ozone depletion. At first only a few environmentalists and atmospheric chemists had paid attention, but Rowland and Molina continued to make new discoveries and to refine their predictions. In 1984 a dramatic hole in the ozone layer was discovered above Antarctica, just as the scientists forecasted.

That got things moving. In 1987, 27 nations signed the Montreal Protocol, the first global environmental agreement to regulate CFCs. And as the science has improved, more and more nations have signed on. At last count there were 183. And each time they meet they strengthen its language and requirements. United Nations Secretary General Kofi Annan has called the Montreal Protocol "perhaps the single most successful international agreement to date." The impact is concrete: Since 1987 the levels of the most critical CFCs and related compounds have stabilized or declined. And while the ozone layer's full recovery will take longer than we initially thought, our efforts to date represent a significant step forward.

Controlling greenhouse gases will be more difficult, because carbon dioxide—the major greenhouse gas—is more closely linked to the global economy than CFCs ever were. Weaning our industries—and changing our personal habits—will be a challenge. But our experience with the ozone layer shows us that the people of the world actually can work together to repair some of our own mistakes, despite our often conflicting political and economic interests. Today, as the CO_2 crisis unites us, we must remember the lesson of the CFC battle: that cool heads can prevail and alter the course of environmental change for the better.

Now it is up to us to use our democracy and our God-given ability to reason with one another about our future and make moral choices to change the policies and behaviors that would, if continued, leave a degraded, diminished, and hostile planet for our children and grandchildren—and for humankind.

We must choose instead to make the 21st century a time of renewal. By seizing the opportunity that is bound up in this crisis, we can unleash the creativity, innovation, and inspiration that are just as much a part of our human birthright as our vulnerability to greed and pettiness. The choice is ours. The responsibility is ours. The future is ours.

VIEW OF STAR-FORMING REGION
S106 IRS4 AS SEEN FROM THE
SUBARU TELESCOPE, MAUNA
KEA, HI, 2001

One of the robotic spacecrafts that America launched years ago to explore the universe took a picture as it was leaving Earth's gravity, a picture of our planet spinning slowly in the void. Years later, when the same spacecraft had traveled 4 billion miles beyond our solar system, the late Carl Sagan suggested that NASA send a signal instructing the craft to turn its cameras toward the Earth again, and from that unimaginable distance, take another photograph of Earth. This is the picture it sent back to us. The pale blue dot, visible in the center of the band of light at the right, is us.

Sagan called it a pale blue dot and noted that everything that has ever happened in all of human history has happened on that tiny pixel. All the triumphs and tragedies. All the wars. All the famines. All the major advances.

It is our only home.

And that is what is at stake. Our ability to live on planet Earth—to have a future as a civilization.

I believe this is a moral issue.

It is our time to rise again to secure our future.

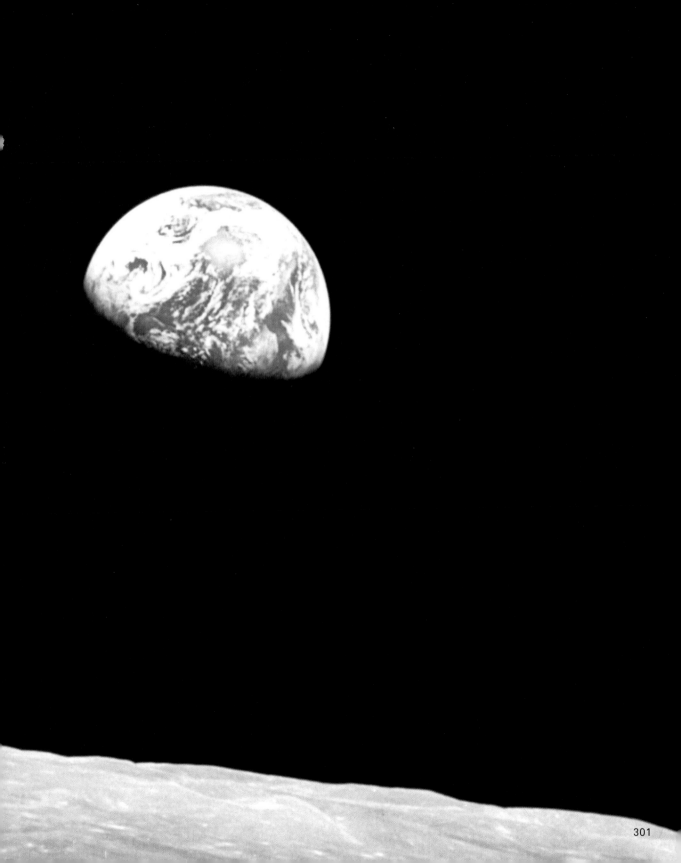

301

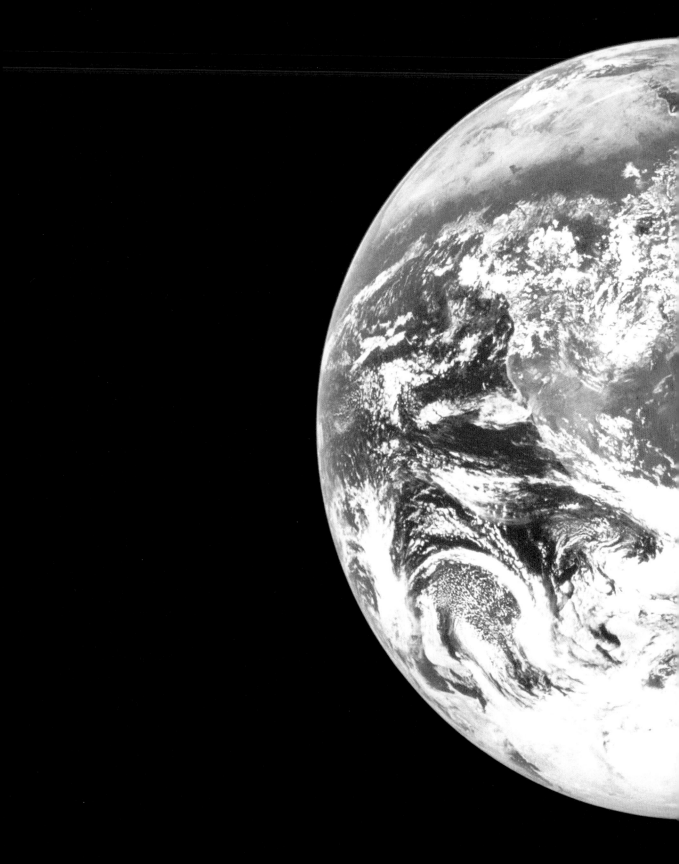

So here's what you personally can do to help solve the climate crisis:

When considering a problem as vast as global warming, it's easy to feel overwhelmed and powerless—skeptical that individual efforts can really have an impact. But we need to resist that response, because this crisis will get resolved only if we as individuals take responsibility for it. By educating ourselves and others, by doing our part to minimize our use and waste of resources, by becoming more politically active and demanding change—in these ways and many others, each of us can make a difference.

On the following pages you will find a range of practical steps anyone can take to reduce the stress our high-tech lives exert on the natural world. As we incorporate these suggestions into our lives, we may well find that not only are we contributing to a global solution, we are also making our lives better. In some cases, the returns are quantifiable: Using less electricity and fuel, for example, saves money. Further: more walking and biking improve our health; diets of locally grown produce bring enhanced taste and nutrition; breathing cleaner air is energizing and healing; and creating a world of restored natural balance ensures a future for our children and grandchildren.

One way to begin making a difference is to learn how the way we live our lives impacts our global environment. All of us contribute to climate change through the daily choices we make—from the energy we use at home to the cars and other vehicles we drive, from the products and services we consume to the trail of waste we leave behind. The average American is responsible for about 15,000 pounds of carbon dioxide emissions each year. This per capita number is greater than that of any other industrialized country. In fact, the United States—a country with 5% of the world's people—produces nearly 25% of the world's total greenhouse-gas emissions.

To calculate your impact on the climate in terms of the total amount of greenhouse gases you produce, visit **www.climatecrisis.net**. There, with the help of an interactive energy meter, you can calculate what your individual impact—your "carbon footprint"—currently is. This tool will also help you evaluate which areas of your life produce the most emissions. Armed with this information, you can begin to take effective action and work toward living a carbon-neutral life.

Save energy at home

Reduce emissions from your home energy use

For most Americans, the easiest and most immediate opportunities to reduce emissions can be found right in their own homes. Most greenhouse-gas emissions that originate in the home are a product of the fossil fuels burned to generate electricity and heat. Many things can be done to reduce these emissions. With an action as simple as changing a few light bulbs, you can take the first important step toward reducing your energy consumption.

Saving energy is not only a good thing to do for the climate crisis: It can also translate to real cost savings. Choosing energy-efficient alternatives for the home can help families cut their energy bills by as much as a one-third, while reducing greenhouse-gas emissions by a similar amount. While many actions can be taken at little or no cost, others may require a small investment up front that will pay for itself in reduced energy bills. Here are some specific ways you can conserve energy in your home.

Choose energy-efficient lighting

Lighting accounts for one-fifth of all the electricity consumed in the United States. One of the easiest and most cost-effective ways to reduce your energy use, energy costs, and greenhouse-gas emissions is to replace the regular incandescent light bulbs in your home with super-efficient compact fluorescent lights (CFLs). CFLs fit right into most regular household fixtures and give off the same warm light, but they are much more energy efficient.

The conventional incandescent bulbs most consumers use in their homes are highly energy inefficient. Only 10% of the energy they consume actually generates illumination, while 90% of it is lost in the form of heat. While CFL bulbs cost more up front, they last up to 10,000 hours—10 times longer than incandescent bulbs—and use 66% less energy.

If every household in the United States substituted even one conventional light bulb with a CFL bulb, it would have the same effect on pollution levels as removing a million cars from the nation's roads.

▶ To purchase CFLs online, visit www.efi.org or www.nolico.com/saveenergy/

Choose energy-efficient appliances when making new purchases

One of the most significant opportunities consumers have to improve home energy efficiency is in the selection of new major appliances such as air conditioners, furnaces, water heaters, and refrigerators. Choosing models that have been designed to use energy efficiently will save you money over time and reduce greenhouse-gas emissions.

The U.S. Environmental Protection Agency's Energy Star Program Web site provides useful information to help with consumer decisions.

▶ For information about the newest energy-efficient appliances, visit www.energystar.gov/products

Properly operate and maintain your appliances

While buying energy-efficient appliances is a good first step in helping to reduce your long-term greenhouse-gas emissions, you can also improve the energy efficiency of older appliances. For example, refrigerators should not be placed next to heat sources such as ovens, dishwashers, and heaters that force them to overwork in order to maintain cool temperatures. A refrigerator's condenser coils should be kept dust-free to ensure unblocked airflow through the unit's heat exchanger. Any appliance's filters should be regularly cleaned or changed.

Another energy saving tip: Instead of running frequent partial loads in your dishwasher or washing machine, save energy by running only full loads. When you have time, wash your dishes by hand, and use a clothesline to dry your clothes instead of the dryer.

▶ The American Council for an Energy Efficient Economy has a checklist available to guide home energy savings, including how to operate appliances. It also has an extensive FAQ section and offers a book you can order for a fuller explanation. Visit http://aceee.org/consumerguide/chklst.htm For additional tips, visit http://eartheasy.com/live_energyeffic_appl.htm

Heat and cool your house efficiently

Heating and cooling your house can be a major energy drain, typically accounting for about 45% of a household's total energy use. Pay attention to how high or low your thermostat is set, avoiding unnecessary cooling or heating. Lowering your heat by just a few degrees in the winter and setting your air conditioner a couple of degrees higher in the summer can add up to real energy savings over time. And using a programmable thermostat allows you to adjust temperatures automatically—for example, while you are sleeping or at work. Also, where possible, install "smart meters" and explore combined heat and power systems.

Insulate your house

Properly insulating your house can save money by eliminating energy leaks that add to your heating or cooling needs. A drafty house lets warm air escape in the winter and lets cool air out in the summer, which puts more pressure on heating and cooling systems—and expends more energy—to keep the house comfortable.

Check for drafts around windows and doors and seal up any leaks, or consider installing higher-efficiency windows. Make sure to seal all attic vents and ducts. Insulate your water heater and hot-water pipes to help keep the heat in the water, where it belongs.

▶ **For more specific information, visit www.simplyinsulate.com**

▶ **The Consumer Federation of America's checklist of Ten Simple Ways to Cut Energy Costs includes these and other suggestions to reduce your greenhouse-gas emissions. Visit www.buyenergyefficient.org**

Get a home energy audit

Performing a comprehensive energy audit can help identify areas of your home that are consuming the most power. For an informative do-it-yourself tool to conduct your own audit, visit **www.energyguide.com**. This site will take you through a step-by-step evaluation of your home, factoring in the structure of your house or apartment, number of rooms, type of heating system, etc. Using this data, the guide offers individualized suggestions for how to reduce energy use while providing tools to calculate the amount of potential energy savings from specific actions. The typical household spends an average of $1,500 per year on energy and can save as much as $450 or more by implementing some simple energy-efficiency measures. There are also professional home energy auditors who can give you thorough home energy-efficiency assessments.

▶ **To find an energy specialist in your area, contact your utility company or state energy office, or visit www.natresnet.org/directory/ rater_directory.asp#Search**

Conserve hot water

Heating water is one of the major draws on household energy. You can cut energy use by setting your water temperature no higher than 120°F.

The 10 most common misconceptions about global warming

MISCONCEPTION 1

"Scientists disagree about whether humans are causing the Earth's climate to change."

In fact, there is strong scientific consensus that human activities are changing the Earth's climate. Scientists overwhelmingly agree that the Earth is getting warmer, that this trend is caused by people, and that if we continue to pump greenhouse gases into the atmosphere, the warming will be increasingly harmful.

You can also conserve hot water by taking showers rather than baths, and by installing efficient low-flow shower-heads.

Consider the water requirements of appliances such as dishwashers and washing machines, as some use less hot water than others. For example, front-loading washing machines are much more efficient than top-loading machines. Washing clothes in warm or cold water, rather than hot, can be a big energy saver as well.

Reduce standby power waste

Many appliances—including televisions, DVD players, cell phone chargers, or any other piece of equipment that has a remote control, battery charger, internal memory, AC adapter plug, permanent display, or sensor—use electricity even when they are turned "off." In fact, 25% of the energy a television uses is consumed when it is not even powered on. The only way to be sure your appliance is not using power is to unplug it, or to plug it into a power strip, which you can then switch off. (Power strips do consume a small amount of energy, but far less than

MISCONCEPTION 2

"Lots of things can impact climate— so there's no reason we should single out CO_2 to worry about."

Climate is sensitive to many things besides carbon dioxide—sunspots, for one, as well as water vapor. But this just proves how much we should worry about CO_2 and other human-influenced greenhouse gases. The fact that the climate system has been shown to be sensitive to many sorts of natural changes throughout history should serve as a red flag: We need to pay close attention to the massive and unprecedented changes we're causing. We have become more powerful than any force of nature.

the phantom load appliances leak when plugged in directly.)

▶ For more information on standby energy, visit www.standby.lbl.gov/index.html and www.powerint.com/greenroom/faqs/htm

Improve the efficiency of your home office

Energy-efficient computers are equipped with a power management feature that, when enabled, causes the computer to go into a low-power mode. Because computers are commonly left on when not in use, enabling power management can save 70% of the energy normally used by a computer.

Also be aware that laptop computers are 90% more energy efficient than desktop models. Inkjet printers consume 90% less energy than laser printers, and printing in color uses more energy than printing in black and white. When possible, choose multi-function devices that print, fax, copy, and scan, as they use less energy than individual machines would.

▶ For more information on Energy Star computers, printers, and other office equipment, visit www.energystar.gov/index. cfm?c=ofc_equip.pr_office_equipment

Switch to green power

Although most energy in the United States comes from fossil fuels, more and more people are electing to use energy generated by cleaner sources such as the sun, wind, the heat of the Earth, or the burning of biomass.

▶ **For more information about these various alternative sources of energy, visit www.eere. energy.gov/consumer/renewable_energy**

In fact, wind and solar power are among the fastest-growing sources of energy, both in the United States and around the world.

▶ **For more information about solar energy, visit www.ases.org/ and for wind energy, visit www.awea.org**

There are a number of different ways to participate in this shift to renewable energy. Many homeowners have begun to produce their own electricity by installing solar photovoltaic cells, wind turbines, or geothermal heat pumps. It is estimated that some 150,000 households have become energy self-sufficient, removing themselves entirely from the energy grid. Many more have reduced their reliance on public utilities, using them only to supplement the renewable power they generate themselves.

In some states, households that produce more electricity than they need for their own use can sell the surplus back to the utility. This is called "two-way" or "net" metering. In this manner, individuals can not only reduce their own carbon emissions, but also supply clean energy to the public utility.

▶ **For more information about net metering, visit www.awea.org/faq/netbdef.html**

Many state and local governments, and some utility companies, offer personal tax credits or subsidies for renewable energy projects.

▶ **For more information, visit the Database of State Incentives for Renewable energy at www.dsireusa.org**

For those who aren't in a position to install their own renewable-energy systems, there is another way to participate in the shift to green power. In many regions, consumers can contract with their utility companies to receive energy from more environmentally friendly sources. There may be a slightly higher cost for green power, but in general the premium is negligible and will likely come down as more consumers elect this option.

▶ **For more information, visit www.epa/gov/greenpower or www.eere.energy.gov/greenpower**

If green power is not available through your public utility, you have the option of purchasing Tradeable Renewable energy Certificates (TRCs) to offset your energy use.

▶ **For more information, visit www.green-e.org**

Get around on less

Reduce emissions from cars and other forms of transportation

Almost one-third of the CO_2 produced in the United States comes from cars, trucks, airplanes, and other vehicles that transport us from place to place, or are used in the course of producing and delivering the goods and services we consume. More than 90% of this travel is by automobile, which means that fuel-economy standards are of critical importance. Average gas efficiency for passenger vehicles has actually declined over the last decade, largely because of the increased popularity of SUVs and light trucks. New regulations that impose more stringent standards on these vehicles will hopefully reverse this trend, and further innovations in gas economy, alternative fuels, and hybrid technology will provide more eco-friendly options. Here are some solutions on the horizon, as well as a few things you can do in the meantime to reduce the carbon emissions you produce in the course of your travels.

Reduce the number of miles you drive by walking, biking, carpooling, or taking mass transit wherever possible

The average car in the United States releases about one pound of carbon dioxide for every mile driven. Avoiding just 20 miles of driving per week would eliminate about 1,000 pounds of CO_2 emissions per year.

▶ For advice on how to lobby for better pedestrian conditions, visit www.americawalks.org, and for better biking conditions, visit www.bikeleague.org

▶ A free national service is available to help you coordinate your travels with other commuters. For more information visit www.erideshare.com

▶ For more information about how to use and support the expansion of mass transit, visit www.publictransportation.org

Drive smarter

Some simple changes in driving habits can improve your vehicle's fuel efficiency and reduce your greenhouse-gas emissions when you must drive. Avoid commuting in rush hour, if possible. You'll waste less time sitting in traffic and your vehicle will consume less fuel. Observe the speed limit—and not only for safety reasons: A car's fuel economy drops off sharply at speeds above 55 mph. Avoid unnecessary idling and keep your car in good running order. Regular maintenance improves performance and reduces emissions. And, as much as possible, plan ahead and combine different errands into one trip.

▶ For specific information about maximizing the fuel efficiency of your car, visit www.fueleconomy.gov/feg/driveHabits.shtml

Make your next vehicle purchase a more efficient one

The recent rise in gasoline prices has increased interest in our cars' fuel efficiency. Driving a car that gets more miles to the gallon will not only save you cash at the gas station, it will also reduce your carbon-dioxide emissions from driving. Every gallon of gasoline burned puts about 20 pounds of carbon dioxide into the atmosphere. So a vehicle that gets 25 rather than 20 miles per gallon produces 10 fewer tons of carbon dioxide in its first 100,000 miles. Comfort needn't be sacrificed to gain fuel economy.

▶ You can look up fuel-efficiency estimates for most cars at the U.S. Department of Energy's online Green Vehicle Guide at www.epa.gov/autoemissions or www. fueleconomy.gov

Hybrids

Hybrid cars run on a mix of gasoline and electricity, and because the battery charges as you drive, they never need to be plugged in. Since the electric motor assists the regular combustion engine, hybrids consume far less gas and are much cleaner for the environment. Some hybrid cars get up to 50 miles per gallon. Demand for these vehicles is growing at a feverish rate, and many new models, including sedans, hatchbacks, SUVs, and pickups, are now or will soon be available.

▶ For more information about how hybrids work and to compare models, visit www.hybridcars.com

Alternative fuels

"The fuel of the future is going to come from fruit like that sumac out by the road, or from apples, weeds, sawdust—almost anything. There is fuel in every bit of vegetable matter that can be fermented. There's enough alcohol in one year's yield of an acre of potatoes to drive the machinery necessary to cultivate the fields for a hundred years." Henry Ford spoke these prophetic words in 1925. Some 90 years later we are seeing the application of such innovations, including the use of numerous biofuels derived from renewable plant materials, including corn, wood, and soybeans. The most commonly used renewable fuels today are biodiesel and ethanol.

▶ For more information on these and other alternative fuels, visit the U.S. Department of Energy's Alternative Fuels Data Center at www.afdc.doe.gov/advanced_cgi.shtml

---

MISCONCEPTION 3

"Climate naturally varies over time, so any change we're seeing now is just part of a natural cycle."

Climate does naturally change. By studying tree rings, lake sediments, ice cores, and other natural features that provide a record of past climates, scientists know that changes in climate, including abrupt changes, have occurred throughout history. But these changes all took place with natural variations in carbon dioxide levels that were smaller than the ones we are now causing. Cores taken from deep in the ice of Antarctica show that carbon dioxide levels are higher now than they have been at any time in the last 650,000 years, which means we are outside the realm of natural climate variation. More CO_2 in the atmosphere means warming temperatures.

Fuel-cell vehicles

A hydrogen fuel cell is a device that converts either pure hydrogen or hydrogen-rich fuel directly into energy. Cars powered by fuel cells may be twice as efficient as similarly sized conventional vehicles—or even more, as new technologies advance efficiencies. A fuel-cell vehicle (FCV) that uses pure hydrogen produces no pollutants: only water and heat. FCVs, while exciting, are still several years away from reaching a mass market.

▶ To learn more about fuel-cell technology, visit www.fueleconomy.gov/feg/fuelcell. shtml

Telecommute from home

Another way to reduce the number of miles you drive is by telecommuting. You'll spend less time and energy on the road and be able to devote more attention to business at the same time.

▶ For further information about telecommuting, visit the Telework Coalition at www.telcoa.org

Reduce air travel

Flying is another form of transportation that produces large amounts of carbon dioxide. Reducing air travel even by one or two flights per year can significantly reduce emissions. Take vacations nearer to home, or get there by train, bus, boat, or even car. Buses provide the cheapest and most energy-efficient transportation for long distances, and trains are at least twice as energy efficient as planes. If your airplane travel is for business, consider whether you can telecommute instead. If you must fly, consider buying carbon offsets to compensate for the emissions caused by your air travel.

MISCONCEPTION 4

"The hole in the ozone layer causes global warming."

There is a relationship between climate change and the ozone hole, but this isn't it. The hole in the ozone layer—a part of the upper atmosphere that contains high concentrations of ozone gas and shields the planet from the sun's radiation—is due to man-made chemicals called CFCs, which were banned by an international agreement called the Montreal Protocol. The hole causes extra UV radiation to reach the Earth's surface, but it does not affect the Earth's temperature.

The only connection between the ozone layer and climate change is almost the exact opposite of the myth stated above. Global warming—while not responsible for the ozone hole—could actually slow the natural repairing of the ozone layer. Global warming heats the lower atmosphere but actually cools the stratosphere, which can worsen stratospheric ozone loss.

▶ For assistance in planning green travel and purchasing carbon offsets, visit www.betterworldclub.com/travel/index.htm

Consume less, conserve more

Reduce emissions by consuming less and conserving wisely

In America, we have grown used to an environment of plenty, with an enormous variety of consumer products always available and constant enticement to buy "more," "new," and "improved." This consumer culture has become so intrinsic to our worldview that we've lost sight of the huge toll we are taking on the world around us. By cultivating a new awareness of how our shopping and lifestyle choices impact the environment and directly cause carbon emissions, we can begin to make positive changes to reduce our negative effects. Here are some specific ideas on how we can achieve this.

Consume less

Energy is consumed in the manufacturing and transport of everything you buy, which means there are fossil-fuel emissions at every stage of production. A good way to reduce the amount of energy you use is simply to buy less. Before making a purchase, ask yourself if you really need it. Can you make do with what you already have? Can you borrow or rent? Can you find the item secondhand? More and more Americans are beginning to simplify their lives and choose to reduce consumption.

▶ **For ideas on how to pare down, visit www.newdream.org**

Buy things that last

"Reduce, reuse, and recycle" has become the motto of a growing movement dedicated to producing less waste and reducing emissions by buying less, choosing durable items over disposable ones, repairing rather than discarding, and passing on items that are no longer needed to someone who can make use of them.

▶ **For more information about the three Rs, visit www.epa.gov/msw/reduce.htm**

▶ **To learn how to find a new home for something you no longer need, visit www.freecycle.org**

Pre-cycle—reduce waste before you buy

Discarded packaging materials make up about one-third of the waste clogging our landfills. Vast amounts of natural resources and fossil fuels are consumed each year to produce the paper, plastic, aluminum, glass, and Styrofoam that hold and wrap our purchases. Obviously, some degree of packaging is necessary to transport and protect the products we need, but all too often manufacturers add extraneous wrappers over wrappers and layers of unnecessary plastic. You can let companies know your objection to such excess by boycotting their products. Give preference to those products that use recycled packaging,

"There is nothing we can do about climate change. It's already too late."

This is the worst misconception of all. If "denial ain't just a river in Egypt," despair ain't just a tire in the trunk. There are lots of things we can do—but we need to start now. We can't ignore the causes and impacts of climate change any longer. We need to reduce our use of fossil fuels, through a combination of government initiatives, industry innovation, and individual action. Dozens of things you can do are outlined in this resource guide.

or that don't use excess packaging. When possible, buy in bulk and seek out things that come in refillable glass bottles.

▶ For more ideas about how to pre-cycle, visit www.environmentaldefense.org/article. cfm?contentid=2194

Recycle

Most communities provide facilities for the collection and recycling of paper, glass, steel, aluminum, and plastic. While it does take energy to gather, haul, sort, clean, and reprocess these materials, recycling takes far less energy than does sending recyclables to landfills and creating new paper,

bottles, and cans from raw materials. It has been suggested that if 100,000 people who currently don't recycle began to do so, they would collectively reduce carbon emissions by 42,000 tons per year. As an added benefit, recycling reduces pollution and saves natural resources, including precious trees that absorb carbon dioxide. And in addition to the usual materials, some facilities are equipped to recycle motor oil, tires, coolant, and asphalt shingles, among other products.

▶ To learn about where you can recycle just about anything in your area, visit www.earth911.org/master.asp?s=ls&a= recycle&cat=1 or www.epa.gov/epaoswer/ non-hw/muncpl/recycle.htm

Don't waste paper

Paper manufacturing is the fourth-most energy-intensive industry, not to mention one of the most polluting and destructive to our forests. It takes an entire forest—more than 500,000 trees—to supply Americans with their Sunday newspapers each week. In addition to recycling your used paper, there are things you can do to reduce your overall paper consumption. Limit your use of paper towels and use cloth rags instead. Use cloth napkins instead of disposables. Use both sides of paper whenever possible. And stop unwanted junk mail.

▶ For information about how to remove your name from mailing lists, visit www.newdream.org/junkmail or www. dmaconsumers.org/offmailinglist.html

Bag your groceries and other purchases in a reusable tote

Americans go through 100 billion grocery bags every year. One estimate suggests that Americans use more than 12 million barrels of oil each year just to produce plastic grocery bags that end up in landfills after only one use and then take centuries to decompose. Paper bags are a problem too: To ensure that they are strong enough to hold a full load, most are produced from virgin paper, which requires cutting down trees that absorb carbon dioxide. It is estimated that about 15 million trees are cut down annually to produce the 10 billion paper bags we go through each year in the United States. Make a point to carry a reusable bag with you when you shop, and then when you're asked, "Paper or plastic?" you can say, "Neither."

▶ To purchase reusable bags, learn more bag facts, and find about actions you can take, visit www.reusablebags.com

Compost

When organic waste materials, such as kitchen scraps and raked leaves, are disposed of in the general trash, they end up compacted deep in landfills. Without oxygen to aerate and assist in their natural decomposition, the organic matter ferments and gives off methane, which is the most potent of the greenhouse gases—23 times more potent than carbon dioxide in global-warming terms. Organic materials rotting in landfills account for about one-third of man-made methane emissions in the United States. By contrast, when organic waste is properly composted

in gardens, it produces rich nutri-ents that add energy and food to the soil—and of course also decreases the volume added to our landfills.

▶ For information about how to compost, visit www.epa.gov/compost/index.htm or www.mastercomposter.com

Carry your own refillable bottle for water or other beverages

Instead of buying single-use plastic bottles that require significant energy and resources to produce, buy a reus-able container and fill it up yourself. In addition to the emissions created by producing the bottles themselves, imported water is especially energy inefficient because it has to be trans-ported over long distances. If you're

MISCONCEPTION 6

"Antarctica's ice sheets are growing, so it must not be true that global warming is causing glaciers and sea ice to melt."

Some ice on Antarctica may be growing—though other areas of the continent are clearly melting and a new 2006 study shows that overall the ice is shrinking in Antarctica. Even if some of the ice is getting bigger, not shrinking, this doesn't change the fact that global warming is causing glaciers and sea ice to melt around the world. Globally, more than 85% of glaciers are shrinking. And in any case, localized impacts of climate change don't cancel out the global trends that scientists are observing.

Some people also mistakenly claim (in Michael Crichton's novel *State of Fear*, for instance) that Greenland's ice is growing. In fact, recent satellite data from NASA shows that Greenland's ice cap is shrinking every year, causing sea levels to rise. The loss of that ice doubled from 1996 to 2005. Greenland lost 50 cubic kilometers of ice in 2005 alone.

concerned about the taste or quality of your tap water, consider using an inexpensive water purifier or filter. Also consider buying large bottles of juice or soda and filling your own portable bottle daily. Using your own mug or thermos could also help reduce the 25 billion disposable cups Americans throw away each year.

▶ For more information about the benefits of using refillable beverage containers, visit www.grrn.org/beverage/refillables/index.html

Modify your diet to include less meat

Americans consume almost a quarter of all the beef produced in the world. Aside from health issues associated with eating lots of meat, a high-meat diet translates into a tremendous amount of carbon emissions. It takes far more fossil-fuel energy to produce and transport meat than to deliver equivalent amounts of protein from plant sources.

MISCONCEPTION 7

"Global warming is a good thing, because it will rid us of frigid winters and make plants grow more quickly."

This myth just doesn't seem to die. Because local impacts will vary, it's true that some specific places may experience more pleasant winter weather. But the negative impact of climate change vastly outweighs any local benefits. Take the oceans, for example. Changes to the oceans caused by global warming are already causing massive die-offs of coral reefs, which are crucial sources of food and shelter for creatures at every stage of the ocean food chain, all the way up to us. Melting ice sheets are causing sea levels to rise, and if big ice sheets melt into the ocean, many coastal cities around the world will flood and millions of people will become refugees. These are just some of the consequences of global warming. Other predicted impacts include prolonged periods of drought, more severe flooding, more intense storms, soil erosion, mass species extinction, and human health risks from new diseases. The small number of people who experience better weather may be doing it in a landscape that is nearly unrecognizable.

In addition, much of the world's deforestation is a result of clearing and burning to create more grazing land for livestock. This creates further damage by destroying trees that would otherwise absorb carbon dioxide. Fruits, vegetables, and grains, on the other hand, require 95% less raw materials to produce and, when combined properly, can provide a complete and nutritious diet. If more Americans shifted to a less meat-intensive diet, we could greatly reduce CO_2 emissions and also save vast quantities of water and other precious natural resources.

▶ For more information about cows and global warming, visit www.earthsave.org/globalwarming.htm and www.epa.gov/methane/rlep/faq.html

Buy local

In addition to the environmental impact that comes from manufacturing the product you are buying, the effects on CO_2 emissions from transporting those goods at each and every stage of production must also be calculated. It is estimated that the average meal travels well over 1,200 miles by truck, ship, and/or plane before it reaches your dining room table. Often it takes more calories of fossil-fuel energy to get the meal to the consumer than the meal itself provides in nutritional energy. It is much more carbon efficient to buy food that doesn't have to make such a long journey.

One way to address this is to eat foods that are grown or produced close to where you live. As much as possible, buy from local farmers' markets or from community-supported agriculture cooperatives. By the same token, it makes sense to design your diet as much as possible around foods currently in season in your area, rather than foods that need to be shipped from far-off places.

▶ To learn more about eating local and how to fight global warming with your knife and fork, visit www.climatebiz.com/sections/news_detail.cfm?NewsID=27338

Purchase offsets to neutralize your remaining emissions

So many things we do in our day-to-day lives—driving, cooking, heating our homes, working on our computers—result in greenhouse-gas emissions. It is virtually impossible to eliminate our personal contributions to the climate crisis through reducing emissions alone. You can, however, reduce your impact to the equivalent of zero emissions by purchasing carbon offsets.

When you purchase carbon offsets, you are funding a project that reduces greenhouse-gas emissions elsewhere by, for example, increasing energy efficiency, developing renewable energy, restoring forests, or sequestering carbon in soil.

▶ For more information and links to specific carbon offsetting organizations, visit www.NativeEnergy.com/climatecrisis

MISCONCEPTION 8

"The warming scientists are recording is just the effect of cities trapping heat, rather than anything to do with greenhouse gases."

People who want to deny global warming because it's easier than dealing with it try to argue that what scientists are really observing is just the "urban heat island" effect, meaning that cities tend to trap heat because of all the buildings and asphalt. This is simply wrong. Temperature measurements are generally taken in parks, which are actually cool areas within the urban heat islands. And long-term temperature records showing just rural areas are nearly identical to long-term records that include both rural areas and cities. Most scientific research shows that "urban heat islands" have a negligible effect on the overall warming of the planet.

Be a catalyst for change

Our actions to help solve the climate crisis can extend well beyond the ways we personally reduce our emissions. By continuing to learn about the state of the environment and what is being done about it, we can inform and inspire others to action. We can bring awareness to our neighborhoods, schools, and workplaces, and find ways to implement programs in these and other communities. As citizens of a democracy, we can support candidates who show a record of environmental responsibility, and we can exercise our right to vote for leaders committed to sustainability. We can voice our disapproval when our elected leaders pursue policies detrimental to the environment, and we can lobby in support of programs and actions that advance global cooperation on this issue. As consumers, we can use our purchasing and investing power to send messages of support to corporations and outlets that show integrity and leadership—and messages of intolerance to those that demonstrate negligence and denial.

Learn more about climate change

There are many Web sites that will give you more information about climate change and global warming. A few good places to start are:

▶ www.weathervane.rff.org
www.environet.policy.net
www.climateark.org
www.gcrio.org
www.ucsusa.org/global_warming

▶ For daily press alert headlines, visit www.net.org/warming

Let others know

Share what you've learned with others. Tell your family, your friends, and your colleagues about climate change and what they can do to participate in the solution. If you have the opportunity, speak to a wider audience or write an op-ed piece or a letter to the editor of your local or school newspaper. Share this book or any other resource that will help others understand the importance of this issue.

Encourage your school or business to reduce emissions

You can further extend your positive influence on emissions well beyond your own home by actively and directly encouraging others to take appropriate action. Think about how you might affect others in your workplace, school, place of worship, and elsewhere.

Vote with your dollars

Find out which brands and stores are making efforts to reduce their emissions and to conduct their businesses in an environmentally responsible manner. Support their practices by purchasing their products and shopping in their stores. Make companies that are negligent aware of your objections. Let them know that until they change their energy-inefficient ways, you'll take your business elsewhere.

▶ **For information about the environmental practices and policies of the companies you buy from, visit www.coopamerica.org/programs/responsibleshopper or www.responsibleshopper.org**

Consider the impact of your investments

If you invest, you should consider the impact that your investments have on climate change. Whether you keep your money in a simple savings account at a bank or local credit union, buy stocks, invest in mutual funds for your retirement, or manage your child's college fund, it matters where your money goes.

There are resources for savers and investors that help ensure that money is being invested in companies, products, and projects that responsibly address climate change and other sustainability challenges. Moreover, considering sustainability issues when making investment decisions doesn't mean lower returns on your investments—indeed, there is evidence that it can actually enhance them. Many of the largest investment organizations in the world have endorsed this view.

▶ **You can read some of this research at www.socialinvest.org/areas/research**

▶ **See how you can make a contribution to stopping climate change, support global sustainability, and do well financially by choosing your investments wisely at www.socialinvest.org/Areas/SRIGuide**

▶ **You can read more about their research and approaches at www.unepfi.org and www.ceres.org**

Take political action

Climate change is a global issue, and your personal actions are a critical first step toward reducing greenhouse gases in the United States and throughout the world. For governments, this is fundamentally a political challenge, which means that individuals can make a difference by pressuring their elected representatives to support measures that have a positive impact on the climate crisis.

MISCONCEPTION 9

"Global warming is the result of a meteor that crashed in Siberia in the early 20th century."

This may sound absurd to some of us, but it's a real hypothesis suggested by a Russian scientist. So what's wrong with it? Basically, everything. The impact of a meteor, much like a volcanic eruption, might have immediate effects on climate if it were large enough. But there is no record of warming or cooling during the period after this meteor hit. The effects that would have been produced by the meteor would have involved water vapor, which only stays in the upper atmosphere for a few years at the most. Any effects would have been short-term, and could not be felt this far in the future.

At all levels of government, decisions are routinely made that have the potential to affect greenhouse-gas emissions. Some U.S. cities have agreed to reduce their emissions in line with the amount the entire United States would have committed to had it signed the international Kyoto Protocol, which requires signatory nations to reduce their greenhouse-gas emissions. In fact, as of December 2005, 194 cities representing 40 million Americans had made this pledge as part of the U.S. Mayors Climate Protection Agreement.

▶ For more information, visit www.ci.seattle.wa.us/mayor/climate

Clearly, we must demand an even more dramatic commitment from our government. If we don't express our views loudly and clearly, the corporate special interests who steadfastly oppose mandatory reductions in greenhouse-gas emissions will continue to prevail.

▶ To learn more about where politicians and candidates stand on global warming, visit www.lcv.org/scorecard

▶ Get the facts and make sure your voice is heard!

MISCONCEPTION 10

"Temperatures in some areas aren't increasing, so global warming is a myth."

It is certainly true that the temperature is not rising at every point on the planet. In Michael Crichton's novel *The State of Fear*, characters pass around graphs that show specific places around the world where temperatures are decreasing slightly or remaining the same. The graphs represent real data from real scientists. But while they may be fact, they don't prove the point. Global warming refers to the rise in the average temperature of the entire Earth's surface due to increased levels of greenhouse gases.

Because the climate is an incredibly complex system, the impacts of climate change will not be the same everywhere. Some areas of the globe—such as northern Europe—might actually become colder. But this does not change the fact that overall, the surface temperature of the planet is rising, as are the temperatures of our oceans. The gains have been demonstrated by several types of measurements—including satellite data—that all show the same general results.

Support an environmental group

There are many organizations doing great work to help solve the climate crisis and all of them can use support. Do some research to find out more about each and then get involved. A few to start with are:

▶ Natural Resources Defense Council, at www.nrdc.org/globalwarming/default.asp

▶ Sierra Club, at www.sierraclub.org/globalwarming

▶ Environmental Defense, at www.environmentaldefense.org/issue.cfm?subnav=12&linkID=15

ACKNOWLEDGMENTS

My wife, Tipper, began urging me to write this book several years ago, arguing that public concern and curiosity about global warming had advanced considerably since I published *Earth in the Balance* in early 1992, and that the public interest would be well served by a new kind of book that combined fresh and up-to-date textual analysis with pictures and graphic images that would make the climate crisis more accessible and less forbidding to a wider audience. And as has so often been the case in our 36 years of marriage, she was not only right, but was patiently and persistently right for a considerable period of time before I *realized* she was right. In any case, she then helped me at every stage of the process to make the idea for the book a reality. Without Tipper, needless to say, this book never would have existed.

After I finally completed the text at the end of 2005, the two of us assembled it along with all the pictures and graphics in the proper order and shipped it from our home in Nashville to my agent, Andrew Wylie, in New York City, on New Year's Eve, just before midnight. And Andrew, as usual, knew exactly how to put the manuscript in just the right hands—in a way that made sure that it was given the best chance to be made into the book that you are now holding.

My experience with Rodale has been nothing short of spectacular. Steve Murphy, the CEO, made this project a personal cause, and moved heaven and earth to complete a complex and unusual project in a beautiful way, in record time. I'd also like to thank the Rodale family, whose lifelong commitment to the environment is inspirational, and whose generous support for this project is greatly appreciated.

I am especially grateful to my editor, Leigh Haber, for her indispensable role in shaping this book, editing it with such skill, for her suggestions and creative ideas—and for making the whole process fun from start to finish, even as we were all working at breakneck speed to meet the impossibly tight deadlines. Thanks also to everyone else at Rodale who worked so hard on this project: Liz Perl and her team, Tami Booth Corwin, Caroline Dube, Mike Sudik and his great production team, Andy Carpenter and his dedicated team, and Chris Krogermeier and her staff.

I am also grateful to Leigh for her decision to invite Charlie Melcher and his wonderful and dedicated colleagues at Melcher Media and mgmt. design to become part of the extraordinary creative team that Rodale organized and that Leigh headed. A special thank you for many late nights to Jessi Rymill, Alicia Cheng, and Lisa Maione. Thanks also to Bronwyn Barnes, Duncan Bock, Jessica Brackman, David Brown, Nick Carbonaro, Stephanie Church, Bonnie Eldon, Rachel Griffin, Eleanor Kung, Kyle Martin, Patrick Moos, Erik Ness, Abigail Pogrebin, Lia Ronnen, Hillary Rosner, Alex Tart, Shoshana Thaler, and Matt Wolf. Charlie and his group have brought an extremely creative approach and a truly impressive work ethic to designing and producing this complex presentation.

In addition, I would like to thank Mike Feldman and his colleagues at the Glover Park Group for their help.

The book and the movie have been separate projects, but the movie team deserves special thanks for all of the many things they have done to facilitate the success of this book, even as the movie was in its final stages of preparation. Thanks especially to:

Lawrence Bender
Scott Z. Burns
Lesley Chilcott
Megan Colligan
Laurie David
Davis Guggenheim
Jonathan Lesher
Jeff Skoll

Special thanks to Matt Groening.

My friend Melissa Etheridge was incredibly responsive and helpful in composing and singing an original song for the end of the movie.

And many years before there was a movie, Gary Allison and Peter Knight helped to organize an early project that turned out to be invaluable in the projects I have pursued over the past couple of years.

Thanks to Ross Gelbspan for his dedication and tirelessness.

The Gore family, at the wedding of Kristin Gore and Paul Cusack, 2005
BACK ROW, LEFT TO RIGHT: Drew Schiff, Frank Hunger, Albert Gore, Al Gore, and Paul Cusack; FRONT ROW, LEFT TO RIGHT: Sarah Gore, Karenna Gore Schiff, Wyatt Schiff (6 years old), Tipper Gore, Anna Schiff (4 years old), and Kristin Gore

Gail Buckland has been a terrific help in finding pictures. She is literally the most knowledgeable person in the world where photo archives are concerned, and I always enjoy working with her and learning from her.

In addition, the people at Getty Images went above and beyond the call of duty to help with this project.

Thanks are due especially to Jill Martin and Ryan Orcutt at Duarte Design—and also Ted Boda, whose place has been taken by Ryan—for all of their countless hours over the past several years helping me find images and design graphics to illustrate complicated concepts and phenomena.

Tom Van Sant has dedicated many years of his life to conceiving and painstakingly creating one of the most remarkable sets of photographic images of the Earth ever made. His images inspired me 17 years ago, when I first saw them, and he has continued to improve them ever since. I am grateful to be able to use the latest one-meter-resolution imagery Tom has produced.

Among the many scientists who have helped me over the years to better understand these issues, I want to single out a small group that has played a particular role in advising me on this book, and the movie that is part of the overall project:

James Baker
Rosina Bierbaum
Eric Chivian
Paul Epstein
Jim Hansen
Henry Kelly
James McCarthy
Mario Molina
Michael Oppenheimer
David Sandalow
Ellen & Lonnie Thompson
Yao Tandong

In addition, three distinguished scientists whose work and inspiration were central to this book are now deceased:

Charles David Keeling
Roger Revelle
Carl Sagan

I am grateful to Steve Jobs and my friends at Apple Computer, Inc. (I am on the Board of Directors) for helping with the Keynote II software program that I have used extensively in putting this book together.

I am particularly thankful to my partners and colleagues at Generation Investment Management for help in analyzing a number of complex questions dealt with in the book. And I want to thank my colleagues at Current TV for their help in locating several images that are used in the book.

I would also like to acknowledge MDA Federal, Inc., for their help in calculating and portraying the imagery to demonstrate with scientific precision the impact of sea level rise in various cities around the world.

Throughout my work on this book, Josh Cherwin, of my staff, has been unbelievably helpful in countless ways. Also, the rest of my entire staff has contributed a tremendous amount:

Lisa Berg
Dwayne Kemp
Melinda Medlin
Roy Neel
Kalee Kreider

Several members of my family played a direct role in helping me with this project:

Karenna Gore Schiff and Drew Schiff
Kristin Gore and Paul Cusack
Sarah Gore
Albert Gore, III
and my brother-in-law, Frank Hunger

All of them have been my constant inspiration and the principal way that I personally connect to the future.

CREDITS

Illustrations by Michael Fornalski
Information graphics by mgmt. design

The publisher and packager wish to recognize the following individuals and organizations for contributing photographs and images to this project:

Animals Animals; ArcticNet; Yann Arthus-Bertrand (www.yannarthusbertrand.com); Buck/Renewable Films; Tracey Dixon; Getty Images; Kenneth E. Gibson; Tipper Gore; Paul Grabbhorn; Frans Lanting (www.lanting.com); Eric Lee; Mark Lynas; Dr. Jim McCarthy; Bruno Messerli; Carl Page; W. T. Pfeffer; Karen Robinson; Vladimir Romanovsky; Lonnie Thompson; and Tom Van Sant

Images are referenced by page number. All photographs and illustrations copyright © by their respective sources.

Inside front cover: Eric Lee/Renewable Films (Al Gore) and NASA (Earth); pages 2–3: Tipper Gore; 6: courtesy of the Gore family; 12–13: NASA; 14: NASA; 16–17: Tom Van Sant/GeoSphere Project; 18–19, gatefold: Tom Van Sant/GeoSphere Project and Michael Fornalski; 22–23: Getty Images; 24–25: Steve Cole/Getty Images; 26–27: Tom Van Sant/GeoSphere Project and Michael Fornalski; 28–29: Derek Trask/Corbis; 32–33: Tom Van Sant/GeoSphere Project; 34–35: Tom Van Sant/GeoSphere Project and Michael Fornalski; 38–39: Antony Di Gesu/San Diego Historical Society; 40: Lou Jacobs, Jr./Scripps Institution of Oceanography Archives/University of California, San Diego; 41: (top) Bob Glasheen/The Regents of the University of California/Mandeville Special Collections Library, UCSD; (bottom) SIO Archives/UCSD; 42–43: Bruno Messerli; 44: Carl Page; 45: Lonnie Thompson; 46–47: U.S. Geological Survey; 48–49: Daniel Garcia/AFP/Getty Images; 51: (photograph) R.M. Krimmel/USGS; (graphic) W. T. Pfeffer/INSTAAR/University of Colorado; 52–53: Lonnie Thompson; 54–55: (composite) Daniel Beltra/ZUMA Press/Copyright by Greenpeace; 56–57: (all photographs) Copyright by Sammlung Gesellschaft fuer oekologische Forschung, Munich, Germany; 58–59: Map Resources; 60–61: (all images) Lonnie Thompson; 62: Lonnie Thompson; 65: Vin Morgan/AFP/Getty Images; 68–69: Tipper Gore; 70: (top) Bob Squier; (bottom) Tipper Gore; 71: (all photographs) Tipper Gore; 74–75: Michaela Rehle/Reuters; 80–81: NOAA; 82: NASA; 85: NASA; 86–87: Don Farrall/Getty Images; 88: Andrew Winning/Reuters/Corbis; 90: Robert M. Reed/USCG via Getty Images; 91: Stan Honda/AFP/Getty Images; 94–95: NASA; 96: (top) David Portnoy/Getty Images; (bottom) Robyn Beck/AFP/Getty Images; 97: (top) Marko Georgiev/Getty Images; (bottom) Reuters/Jason Reed; 98–99: Vincent Laforet/The New York Times; 103: Reuters/Carlos Barria; 104–105: (composite) NASA/NOAA/Plymouth State Weather Center; 107: Reuters/Pascal Lauener; 108–109: Keystone/Sigi Tischler; 110–111: Sebastian D'Souza/AFP/Getty Images; 112: Reuters/China Newsphoto; 113: China Photos/Getty Images; 114–115: Tom Van Sant/GeoSphere Project and Michael Fornalski; 116: (all) NASA; 117: Stephane De Sakutin/AFP/Getty Images; 118–119: Yann Arthus–Bertrand [Road interrupted by a sand dune, Nile Valley, Egypt (25°24' N, 30°26' E). Grains of sand, deriving from ancient river or lake alluvial deposits accumulated in ground recesses and sifted by thousands of years of wind and storm, pile up in front of obstacles and thus create dunes. These cover nearly a third of the Sahara, and the highest, in linear form, can attain a height of almost 1,000 feet (300 m). Barchans are mobile, crescent-shaped dunes that move in the direction of the prevailing wind at rates as high as 33 feet (10 m) per year, sometimes even covering infrastructures such as this road in the Nile Valley. Deserts have existed throughout the history of our planet, constantly evolving for hundreds of millions of years in response to climatic changes and continental drift. Twenty thousand years ago forest and prairie covered the mountains in the center of the Sahara; cave paintings have been discovered there that depict elephants, rhinoceros, and giraffes, testifying to their presence in this region about 8,000 years ago. Human activity, notably the overexploitation of the semi-arid area's vegetation bordering the deserts, also plays a role in desertification.]; 120: Paul S. Howell/Getty Images; 121: (graphic) Geophysical Fluid Dynamics Laboratory/NOAA; 122–123: Tipper Gore; 124: (left) courtesy of the Gore family; (right) Washingtonian Collection/Library of Congress; 125: (all photographs) Ollie Atkins/Saturday Evening Post; 127: Tom Van Sant/GeoSphere Project and Michael Fornalski; 128–129: Derek Mueller and Warwick Vincent/Laval University/ArcticNet; 130–131: Peter Essick/Aurora/Getty Images; 132: (top) Vladimir Romanovsky/Geophysical Institute/UAF; (bottom) Mark Lynas; 133: (graphic) Arctic Climate Impact Assessment; 134: (top) Bryan & Cherry Alexander Photography; (bottom) Paul Grabbhorn; 136–137: Karen Robinson; 139: David Hume Kennerly/Getty Images; 140: (top) New Republic; (bottom) Tipper Gore; 141: White House Official Photo; 142: Naval Historical Foundation; 145: Michael Fornalski; 146–147: Tracey Dixon; 148: Tom Van Sant/GeoSphere Project and mgmt. design; 150–151, gatefold: Tom Van Sant/GeoSphere Project and Michael Fornalski; 153: Benelux Press/Getty Images; 155: Kenneth E. Gibson/USDA Forest Service/www.forestryimages.org; 156–157: Peter Essick/Aurora/Getty Images; 158–160: Nancy Rhoda; 161: (top) Nancy Rhoda; (bottom) courtesy of the Gore family; 162: (left to right, top to bottom) Juan Manuel Renjifo/Animals Animals; David Haring/OSF/Animals Animals; Rick Price Survival/OSF/Animals Animals; Juergen and Christine Sohns/Animals Animals; Johnny Johnson/Animals Animals; Frans Lanting; Michael Fogden/OSF/Animals Animals; Johnny Johnson/Animals Animals; Raymond Mendez/Animals Animals; Leonard Rue/Animals Animals; Frans Lanting; Frans Lanting; Peter Weimann/Animals Animals; Don Enger/Animals Animals; Erwin and Peggy

Bauer/Animals Animals; Frans Lanting; 165: Paul Nicklen/ National Geographic/Getty Images; 166–167: Bill Curtsinger/National Geographic/Getty Images; 168: David Wrobel/Getty Images; 169: (graphic) USGCRP; 170: Janerik Henriksson/SCANPIX/Retna Ltd.; 171: (top) Kustbevakningsflyget/SCANPIX/Retna Ltd.; (bottom) Kustbevakningen/SCANPIX/Retna Ltd.; 174: (all images) Centers for Disease Control and Prevention; 177: Tom Van Sant/GeoSphere Project; 178–179: Frans Lanting; (inset) Jérôme Maison/Bonne Pioche/A film by Luc Jacquet/ Produced by Bonne Pioche productions; 180: British Antarctic Survey; 182–183: (all satellite images) NASA; 184–185: Frans Lanting; 186–187: Mark Lynas; 188: Andrew Ward/Life File/Getty Images; 192: (left to right) Dr. Jim McCarthy; (graphic) Buck/Renewable Films and NASA; 193: Roger Braithwaite/Peter Arnold; 194–195: (graphic) Renewable Films/ACIA; 198–202: (all images) MDA Federal Inc. and Brian Fisher/Renewable Films; 203: Ooms Avenhorn Groep bv; 204–209: (all images) MDA Federal Inc. and Brian Fisher/Renewable Films; 211–213: courtesy of the Gore family; 214–215: Yann Arthus-Bertrand [Refuse dump in Mexico City, Mexico (19°24' N, 99°01' W). Household refuse is piling up on all continents and poses a critical problem for major urban centers, like the problem of air pollution resulting from vehicular traffic and industrial pollutants. With some 21 million residents, Mexico City produces nearly 20,000 tons of household refuse a day. As in many countries, half of this debris is sent to open dumps. The volume of refuse is increasing on our planet along with population growth and, in particular, economic growth. Thus, an American produces more than 1,500 pounds (700 kg) of domestic refuse each year, about four times more than a resident of a developing country and twice as much as a Mexican. The volume of debris per capita in industrialized nations has tripled in the past 20 years. Recycling, reuse, and reduction of packaging materials are potential solutions to the pollution problems caused by dumping and incineration, which still account for 41% and 44%, respectively, of the annual volume of household garbage in France.]; 218–219: Yann Arthus-Bertrand [Shinjuku district of Tokyo, Japan (35°42' N, 139°46' E). In 1868 Edo, originally a fishing village built in the middle of a swamp, became Tokyo, the capital of the East. The city was devastated by an earthquake in 1923 and by bombing in 1945, both times to be reborn from the ashes. Extending over 43 miles (70 km) and holding a population of 28 million, the megalopolis of Tokyo (including surrounding areas such as Yokohama, Kawasaki, and Chiba) is today the largest metropolitan region in the world. It was not built according to an inclusive urban design and thus contains several centers, from which radiate different districts. Shinjuku, the business district, is predominantly made up of an impressive group of administrative buildings, including the city hall, a 798-foot-high (243 m) structure that was modeled after the cathedral of Notre Dame in Paris. In 1800 only London had more than 1 million inhabitants; today 326 urban areas have reached that number, including 180 in developing countries and 16 megalopolises that have populations of more than 10 million. Urbanization has led to a tripling of the population living in cities since 1950.]; 220: Peter Essick/Aurora/Getty Images; 221: Kevin Schafer/Corbis; 222–223: National Geographic; 224–225: United Nations Environmental Programme; 226–227: Stephen Ferry/Liaison/Getty Images; 228: Philippe Colombi/Getty Images; 230–231: NASA; 233: Michael Dunning/Getty Images; 234: (right to left, top to bottom) Bridgeman Art Library/Getty Images; Hulton Archive/Getty Images; Palma Collection/Getty Images; Bettmann/Corbis; 235: Corbis; 236: Dean Conger/ Corbis; 237: Photodisc/Getty Images; 238–239: Beth Wald/ Aurora/Getty Images; 240–241: Baron Wolman/Getty Images; 242–243: USGS; 244–245: David Turnley/Corbis; 246–247: Digital Vision/Getty Images; 248–249: NASA; 257. courtesy of the Gore family; 258: Ollie Atkins/Saturday Evening Post; 259: (top) Ollie Atkins/Saturday Evening Post; (bottom) Ollie Atkins/Saturday Evening Post; 264: The New York Times; 271: White House; 277: (left to right, top to bottom) William Thomas Cain/Getty Images; Koichi Kamoshida/Getty Images; Mark Segal/Getty Images; City of Chicago; Joe Raedle/Getty Images; David Paul Morris/ Getty Images; James Davis/Eye Ubiquitous/Corbis; 278–279: Yann Arthus-Bertrand [Middelgrunden offshore wind farm, near Copenhagen, Denmark (55°40' N, 12°38' E). Since late 2000, one of the largest offshore wind farms to date has stood in the Øresund strait, which separates Denmark from Sweden. Its 20 turbines, each equipped with a rotor 250 feet (76 m) in diameter, standing 210 feet (64 m) above the water, form an arc with a length of 2.1 miles (3.4 km). With 40 megawatts of power, the farm produces 89,000 mW annually (about 3% of the electricity consumption of Copenhagen). By 2030 Denmark plans to satisfy 40% of its electricity needs by means of wind energy (as opposed to 13% in 2001). Although renewable forms of energy still make up less than 2% of the primary energy used worldwide, the ecological advantages are attracting great interest. Thanks to technical progress, which has reduced the noise created by wind farms (installed about one-third of a mile, or 500 m, from residential areas), resistance is fading. And with a 30% average annual growth rate in the past four years, the wind farm seems to be here to stay.]; 284–285: courtesy National Archives; 287: Callie Shell; 290–291: (left to right, top to bottom) Hulton Archive/Getty Images; G.A. Russell/ Corbis; National Archives; Time Life Pictures/U.S. Coast Guard/Time Life Pictures/Getty Images; Bettman/Corbis; AFP/Getty Images; 292–293: NASA; 295: NASA; 296–297: Subaru Telescope, National Astronomical Observatory of Japan. All rights reserved; 298–299: NASA; 300–301: NASA; 302–303: NASA; 306: Royalty-Free/Corbis; 311: Paul Costello/Getty Images; 314: Michael S. Yamashita/ Corbis; 319: Joel W. Rogers/Corbis; 323: Tipper Gore; inside back flap: National Optical Astronomy Observatory/ Association of Universities for Research in Astronomy/ National Science Foundation

CANEY FORK RIVER,
CARTHAGE, TN, 2006.
PHOTOGRAPH BY TIPPER GORE

This book is printed on Appleton Green Power Utopia 80# matte. The paper is made of 30% postconsumer waste. The virgin fibers are grown under a certified forestry management system. The pulp is produced chlorine-free using 100% green power.

This is the first book produced to offset 100% of the carbon emissions generated from production activities with renewable energy. By helping build new Native American and Alaska Native wind farms and a new family dairy farm methane energy project through NativeEnergy, this publication is carbon neutral. Rodale is helping to finance these important projects by supporting a share of the CO_2 offsets they are estimated to produce over their operating lives. For more information and to offset your own carbon footprint, visit www.nativeenergy.com.